Wool Away

Her Majesty the Queen and the Duke of Edinburgh at the Royal Command Shearing Performance at Napier in January 1954. The author is answering a question from the Queen about the breed of sheep shorn. On the right of the photograph is Mr. Ivan Bowen, equally eminent as a shearer, who partnered his brother in the shearing demonstration.

Wool Away

THE ART AND TECHNIQUE
OF SHEARING

Godfrey Bowen

INTRODUCTION BY ALLEN FANNIN

Van Nostrand Reinhold Company, New York

Whitcombe & Tombs, New Zealand

To My Father

Van Nostrand Reinhold Company Regional Offices:
New York   Cincinnati   Chicago   Millbrae   Dallas

Van Nostrand Reinhold Company International Offices:
London   Toronto   Melbourne

Copyright © 1955 by Whitcombe and Tombs Ltd.

Introduction Copyright © 1974 by Allen Fannin

Library of Congress Catalog Card Number 73-16702

ISBN 0-442-20964-9   Van Nostrand Reinhold

ISBN 0-7233-0015-1   Whitcombe & Tombs

Published in 1974 by Van Nostrand Reinhold Company
A Division of Litton Educational Publishing, Inc.
450 West 33rd Street, New York, N.Y. 10001

Published in New Zealand and Australia
by Whitcombe & Tombs, Publishers,
A Division of Whitcoulls Ltd.

1 3 5 7 9 11 13 15 16 14 12 10 8 6 4 2

Bowen, Godfrey.
    Wool away!
    Reprint of the ed. published by Whitcombe and Tombs,
Christchurch, N. Z.
    Includes bibliographical references.
    1. Sheep-shearing.   I. Title.
SF379.B67   1974   636.3'1'4   73-16702
ISBN 0-442-20964-9

Contents

Queen Elizabeth and the Duke of Edinburgh renew acquaintance with the author (right) and his brother Ivan, who gave a shearing demonstration for the Royal couple during their 1963 visit to New Zealand.

Foreword

In preparing this textbook on shearing I have not written with the intention of setting myself up as the only authority on the subject, but have written what I believe to be right, based on my own experience. I feel there is a real need for a complete book on shearing, bearing in mind that there are 714 million sheep in the world, producing 3831 million pounds of wool, and that there is only one way to take the wool off their backs —i.e. by a man. So I trust this book will be a help and guide to those who read it.

Among the thousands of shearers that have toiled over this fascinating work in past years, many champions have stood out, all with their own particular styles. For this reason techniques and styles of shearing have always been controversial and much-debated topics. There will be differing opinions about the principles this book lays down, but I submit this style and technique with the firm belief that it is easy on the man, simple to learn, in practice gives maximum quality of work, and, when fully mastered, gives top-line speed.

I trust it will be read in the spirit in which it is written and that the old shearing maxim will still hold true: 'One good man will never condemn another.'

G.B.

Introduction

Since its first publication in New Zealand in 1955, *Wool Away*
has not received its due share of recognition in most parts of the
United States. The changes that have taken place in the American wool industry in the span of the nearly 20 years between
1955 and this new printing have made the book even more
significant than when it first appeared. In addition to the scarcity
of commercial shearers, a growing number of craftsmen, including weavers and hand spinners, have become interested in
trying their hands at shearing on the small flocks they have
access to.

Wool growing in the United States

We might, for the uninitiated reader, survey the domestic
wool growing scene to gain a slight perspective on the past 20
years. Basically, wool growing in the U.S. is divided into a number of categories, the two principal divisions being the range
flocks and the farm flocks. In the range flocks, large bands of
several thousand head are turned out on vast acreages of natural, unimproved pasture land.

As a rule, range flocks form the primary business of any
given enterprise. In the second instance, farm flocks, sheep are
kept on improved pasture or even crop residue and generally
number only a few hundred head. Often, farm flocks are a sub-
sidiary operation to beef or dairy. Range flocks are found primarily in the western part of the United States, while farm flocks
predominate in the Mississippi Valley and the East.

Another division in domestic wool growing has to do with
geographical area. The largest and most significant area com-
prises the Territory Wool states, which include Montana, Wy-
oming, Colorado, New Mexico, Idaho, Utah, Arizona, Nevada,
Washington, and Oregon. All these states combined produce

approximately 40% of the total domestic clip. The second geographical division takes in the Fleece Wool states, which are all the remaining states not producing Territory Wools. The Fleece Wool clip is approximately 30% to 32% of the total. Texas and California produce such considerable amounts of shorn wool that they are taken as separate categories in themselves. Texas Wools comprise some 21%, while California yields about 7% of the total.

There is a correlation between the geographical areas and the flock types. In general, most of the Territory Wools are produced from range flocks and the Fleece Wools come from farm flocks. With increasing numbers of sheep being raised for meat, there is often more overlap than previously.

Because these areas differ so widely in geography, they also differ in ways that are significant to shearing practices. Firstly, the Territory Wool states provide the closest approximation here in the United States to the climactic and grazing conditions that prevail in Australia and New Zealand, where Godfrey Bowen and his brother Ivan developed the shearing style shown in *Wool Away*. Flocks of sheep in this area often range over many hundreds of thousands of acres. Because of the size of most of the Territory Wool operations, the sheep are brought to the shearers. Large shearing pens are provided where many animals can be held until sheared. These installations are often permanent, or at least semi-permanent, since two shearings per year are done in some locations, requiring constant, full-time use of the pens.

The Fleece Wool states are most sharply contrasted with the Territory Wool states by the much smaller size of their operations. Geographically, they do not contain vast stretches of grazing land, which in part accounts for the fewer number of sheep per flock. Shearing in the Fleece Wool states is done mainly by shearers who travel from flock to flock instead of shearing in central locations. An exception to this might be when meat lambs are sheared at a packing plant before slaughter.

In all areas of the country, sheep are raised for both meat and wool, with some emphasis on one product or the other in certain places. The picture overall, however, is that less wool is being produced as the primary product, and more of it simply as a by-product of sheep raised for meat. Wool consumption in the United States, for all kinds of end uses, has dropped from what it was 20 years ago, and many of the large New England woolen and worsted mills have closed down. Greater use of man-made and synthetic fibers is the most important reason for this drop. Then too, sheep raising has never been the most profitable of agricultural operations when compared to beef or dairy. In parts of the Territory Wool area, some sheep ranchers are beginning to convert to beef.

Interestingly, however, as of this writing, a sizable portion of the United States wool clip is beginning to go to foreign buyers at record prices. While this condition may not be permanent, nor will it necessarily return domestic wool production to what it has been, it may at least cause many growers to stay with sheep. Another point on the positive side is the increasing world-wide concern with ecological matters. It is a well known fact that practically all synthetic fibers, while superior to wool in some applications, have the disadvantage of having to be produced from a fractional distillation of oil, a depletable raw material. Only rayon and cellulose acetate are not produced from oil. Wool on the other hand, is a renewable resource and, as such, may well be more dependable in an extremely long-range sense.

Shearing for the Amateur and Professional

One might ask, if the domestic production of wool is down from 20 years ago, of what use is an American printing of a book on sheep shearing. The answers are numerous. First of all, according to some of the larger sheep growers in all parts of the country, shearers are in short supply. This is understandable in part because there is no systematic training for shearers here similar to the system maintained by the New Zealand Wool Board. In some parts of the Northeast, shearers are so scarce that there are instances where the shearer would have to charge for the shearing and take the fleece as well to make the traveling worth his time. Such a situation is especially costly for young people, with little or no farm experience who are attempting to raise sheep. Because each of their flocks number only a handful of animals, obtaining the services of a shearer, if one is available, would often involve a mileage charge in addition to the shearing fee.

Secondly, even in areas where permanent, full-time pens are arranged, a text covering what has come to be the most efficient means of shearing, can be useful in training new shearers. Lastly, *Wool Away* is the single most carefully written book on the subject and for it to be made more available to American readers is important in continuing the shearing craftsmanship that Bowen established.

There are, to be sure, those in American sheep circles who would dispute some of what the author has to say. On the other hand, Bowen's writing style, like his shearing, is so completely clear that the reader can see plainly the logical formulation and, hopefully, the validity of his principles.

A number of American shearers, particularly in the East, advocate that the beginner start with the self-contained electric shear, which, unfortunately, Bowen does not mention at all. Other shearers consider this machine little better than an oversized barber's hair clipper. The reason usually given for advocating the self-contained unit for beginners is that it allows more flexibility in following an active, jumpy animal. This flexibility is possible because the only connection between the handpiece and the power source is an electrical wire. The motor is contained in the handpiece.

However, the most efficient and effortless shearing style demands that the shearer remain in one place, turning the sheep from one side to the other only the minimum number of times needed to shear all parts. With a self-contained electric shear, it becomes too easy for a beginner to develop the bad habit of following an animal around instead of taking complete control of the sheep's movements from start to finish. Shearing one sheep is hard work, and 25 head is certainly 25 times as hard. It is clearly demonstrable that, for the least physical strain for the shearer and the sheep, the operation has to be done as quickly as possible, consistent with highest fleece quality. Therefore, any shearing machine that cannot allow for fast work or that encourages the shearer to chase the sheep all over the board, defeats its purpose. The aim, as Bowen continually points out, is to do maximum amount of work with the minimum number of non-productive movements.

The self-contained unit is not to be recommended for serious work for several reasons. First and most important, Bowen's main emphasis, with which we strongly agree, is that the beginner should develop a rhythm and smoothness at the outset. Because the flexibility of the self-contained unit seemingly allows the beginner to move around more, it becomes difficult, if not impossible, for excess movement and effort to be eliminated at a later stage.

Secondly, even if the student of shearing were able to develop an efficient style using the self-contained shear, physically, this machine does not have the power of the drop tube machine which top shearers use, and which is described in this book. The self-contained shear uses a built-in electric motor, similar to that used in sewing machines, portable electric hand tools, etc., and which is known as an AC/DC universal motor. It depends for its power on very high RPM geared down to the working speed required by the tool on which it is used. For certain applications, such as portable electric handsaws, drills, etc., the universal motor is ideal, because a gear box is built onto the motor shaft to increase output torque. However, the motor is a poor compromise for use in a shear, because such a gear reduction on a shear handpiece makes the machine too heavy and cumbersome to use.

A universal motor also has one electrical characteristic that is a detriment to its use in shearing. When a heavy demand is made on the shear, the motor responds by attempting to draw more current, and becomes quite hot in the process. Working for long periods with a hot handpiece can be very uncomfortable. Anyone who has ever worked with electric hand tools knows also that universal motors, because of their high speed, are extremely noisy.

On the whole, the only advantage to the self-contained electric shear, aside from its low initial cost, is in intermittent use such as fitting out show animals, eye wigging of wool-blind sheep, and crutching at lambing and breeding time.

Students of sheep shearing who intend to shear professionally, can make use of most of what Bowen discusses without any modification. Readers who intend to shear only a limited number of animals for themselves, will need to make only few changes. Under no circumstances should this be taken to mean that the actual shearing movements and style has to be altered. Whether shearing one or one thousand head, the wool still comes off the same way. Modifications that are needed would include redesigning of the holding pens to accommodate fewer animals, changing the time of the shearing run, and perhaps eliminating the shearing board with the porthole. For only a few animals, shearing can be done on any clean surface with some provision for hanging the drive unit and motor.

The sheep grower with a small flock, doing his own shearing, will have to avoid making compromises to good craftsmanship. It is a temptation to put off learning all the proper movements simply because one only has a few sheep to shear. If one does not have a large enough flock to provide sufficient practice, then the movements have to be watched even more carefully.

A 75-year-old shearer, with over half a century of shorn sheep behind him (and a lot more to go), once said that even if all the sheep in the world were raised for meat, the wool would still have to come off. It would seem wasteful, therefore, not to make use of the wool, or not to do it in the most efficient way. Despite research into other ways of separating wool from sheep, all indications are that the skill of a good shearer still leads by a long margin as the most efficient way to do the job.

ALLEN FANNIN

CHAPTER ONE

Shearing: Types of Sheep: Quality of Work: Style

The Romance of Shearing

WAS THERE EVER a job quite like this one! Shearing is hard work (probably among the hardest we've got), work that calls for much more than just physical strength and exuberance— rather for balance, grace, rhythm, suppleness, with eye, brain and hand in smart co-ordination. All these are required to make a 'gun' shearer. Shearing demands still another great quality —the ability to work and keep going, mastering with a big heart such things as very high temperature (with an iron shed full of sheep on a hot day, heat is almost unbearable), aching back, grease boils, maybe a touch of biliousness or some such ailment, and yet a good man will not give in or knock off. Yes, it is a Man's job (no-one with a chicken heart should ever consider taking it on), and yet over the years it has developed into a competitive sport and for past years, right back in early blade-shearing history, it was always a race, and an honour to shear for the 'ringer's' stand.

I look back on the many competitions I have shorn in, and I was always pleased with the fact that while training for those competitions—i.e. while shearing—I got paid for my training. In any other sport, even chopping, training is hard work with no pay.

How many times have I heard an old hand say, 'Well, this is my last season'? And yet next year he is at it again. It does get in a man's blood. Over the years in my many travels I have called into numerous sheds (I couldn't pass them if I saw sheep coming out of the porthole) just to have a look, but have usually finished up by taking off my coat and shearing a sheep or two myself. To readers who have not shorn many sheep this may sound a bit odd, but shearers (all you good 'cobbers' that have shorn sheep) know the romance, glory, 'shi-ack' and fascination of this grand job. Good weather, good shed, good sheep, good boss, and a good gang create an atmosphere of work and action, and yet of good-natured repartee and humour that gladdens the heart and spirit of any real man fortunate enough to be associated with it. Shearing has its own language, many of the terms of which I have listed later in this book, as they are too valuable to leave out. I could fill this book with my experiences but this is essentially a textbook.

I remember one place we called the boss the Red Light and the manager the Blue Light. A shearer went in to catch a sheep, saw the boss coming, and he forgot to yell out, remembering only when he had taken the belly off. He yells, 'Red Light's coming boys'. A voice behind him: 'Ah, me boy, the Red Light's here. One "pinks" them a bit when the Red Light's around.'

The shearers squint along the pens; they squint along the chutes;
The shearers squint along the board to catch the Boss's boots;
They have not time to straighten up, they have no time to stare,
But when the Boss is looking on, they like to be aware.

Different Types of Sheep

All shearing differs as sheep differ. Some are slow and awkward shearing, others easy and fast. Climatic conditions, the treatment sheep have had in the months before shearing, and the breeds of the sheep all play a big part in a shearer's tally. For this reason a shearer's tally can fluctuate considerably, depending on the type of sheep he is shearing. It is a mistake for young shearers to make 'tally' paramount in their ambition to progress in shearing when quality of work would have to be the first point taken into account. Quality of work stands in the opinion of all sound-thinking sheepowners as a much more important factor than speed. In short, a clean, steady, reliable shearer is more valuable and more sought after than a fast rough shearer. To analyse it, quality of work coupled with speed gives the top mark in shearing.

As one who has shorn all breeds of sheep, my advice to shearers is that they should not chase the tally too much, for a shearer can only go as fast as the sheep will let him. Fine-wool sheep, especially Merinos, are much slower than crossbred sheep. Wrinkles, dense wool, trimmings, all have a bearing on shearing.

The fact that fine-wool sheep are usually run on hard country is another reason why these sheep are slower to shear. They do not suit good country because their feet go quickly on heavy or wet land. Even here, New Zealand sheep vary a great deal in each district, and, as I have publicly demonstrated when shearing, conditioning myself to the different breeds and types of sheep has been one of my problems. However, with the right mental approach and correct technique no sheep presents any real problem.

The financial return to the shearer is much the same on fine or coarse wool, for fine-wool men pay up to twice as much as for crossbred sheep. In this country, Merinos, Corriedales and Half-breds are classed as the fine-wool sheep, and all the other breeds in the long-strong-wool or mutton class. (See Chapter Twelve, Sheep Breeds.)

Quality of Work

The need for good workmanship in shearing must be emphasized. Figures taken out in this country have shown that the loss to our annual wool clip by second cuts is very large, costing many hundreds of pounds (money). Other wool-producing countries report the same condition in their wool harvest. My advice to all is to concentrate on shearing to a pattern, keeping the comb on the sheep, watching yourself for 'two-cut' and keeping it to a minimum. (It is possible to shear a sheep almost twice.) But a shearer is only paid for the first time.

Cutting sheep is another point that must be mentioned. Should a shearer continually cut sheep? I say, no! A shearer will shear fewer sheep in a day if he is cutting them, as when the comb digs in it makes the sheep kick and the dropper moves around and loses its true swing. If any reader has trouble with cuts he should look to his gear, and the doing up of combs, and I am confident he will adapt a comb that will eliminate cuts. Remember that one hears some shearers called butchers. In fact it is not the shearer that does the butchering, it is the comb on his handpiece. I would point out here that there is a big difference between what is termed a skin scratch and a cut. Shearing sticky sheep is practically impossible without skin scratches (a small mark on the outer skin that does not bleed and is not apparent the next day). This never affected any sheep. Cuts are a different matter, with the skin cut clean through or off, and the comb has run into the flesh of the sheep causing injury, pain, and bleeding.

The two worst cuts of all are ewes' teats cut or cut off, and wethers' pizzles cut or cut off. With such cuts the ewe's breeding value is gone, while wethers can be seriously affected—all by a careless swipe of a shearer's handpiece, or by trying to hurry over these places for a tally. Putting speed before quality, the shearer has cost the sheepowner a considerable yet perhaps unknown figure in money. Yet I believe that if a shearer does continually cut sheep in these places he shears fewer in a day, as he is mentally conscious of his shearing, looking for the 'boss's boots'. He loses a lot more time than if he had taken two or three seconds longer in either place to make sure this did not happen. I recall a verse of an old shearing poem:

> Bogan Bill was shearing rough, chanced to cut a teat,
> He stuck his leg in front at once, and slewed the ewe a bit.

An accident will occasionally happen, when a sheep kicks the dropper, and before the shearer has a chance his handpiece has given a kick. Farmers recognize this point, but shearers, if it does happen, should let the boss know so that the sheep can be marked. He will appreciate it much more than if the sheep is just put out. He would probably see it outside in any case.

It has to be remembered that until recently there was really no place to which a shearer could go to learn to shear, except down the end of the board, and the only pat on the back he gets is when the boss says, 'You better make a better job of it, son': the point is, he doesn't know how to. I believe that organized instruction and courses on shearing, by the right men, will do much to raise the standard of shearing.

Another factor which has a tendency to affect the standard of work is the desire of some shearers to chase big tallies before they are capable of doing them. Personally, I believe no tally should be called a tally unless the finished job on the sheep is good and unless there is the minimum of second cuts in the wool.

A clean shearer is usually rewarded for his efforts: any bonuses received are for a good job well done. I recall a great friend of mine, who told me about his early shearing days in New Zealand when machines had not been going long. They were shearing for a hard Scotsman, and the backs of the sheep were full of sand and dirt. The comb wouldn't go through, so they went high over sand on the back of the sheep, resulting in the sheep having two inches of wool in clumps over their backs. The boss was away this day, and that night a heavy downpour of rain washed all the sand out. When he saw the sheep next morning, he said: 'Um! You could chop it off with a —— axe better'. It is not always the shearer's fault.

Style

There are several known styles of shearing. They range from the old blade method of stretching the sheep and shearing more round and round, to the modern method of dumping or crimping the sheep and following the blows right through more along the body of the sheep; and between the two styles there are various mixtures of both. In fact, on almost any place on a sheep there are at least two ways of doing it.

However, I condense 'style' to the following basic principles which are essential to easy and good shearing.

1 MINIMUM PHYSICAL EFFORT

No style has merit unless it gives a minimum of physical effort. If any blow in shearing, no matter how pretty or flourishing it looks, costs effort, it should be ruled out, as a man must get his sheep easy, and effort or strain expended on any place will gang up on him at the end of the day. There is a temptation when shearing, for a man to flourish and show off, which to the observer may look spectacular, but in reality is not giving the maximum result. Young shearers should never be influenced by this flourish, but should rather take notice of the man who shears in a quiet, business-like way, and, while appearing not to be going so fast or working so hard as his mates who are full of action and flurried movements, yet at the end of the day rings them by quite a few sheep—that shearer has something.

2 CONTROL OF THE SHEEP

The next basic feature in style is control of the sheep. A good shearer can be likened to a good billiards player—he won't make one shot unless the next one is waiting. Have your sheep moving into position the whole time. This is especially important in machine shearing, as position and the lie of the sheep is everything. In controlling sheep watch balance. Sheep and shearer between them make up approximately three hundred pounds weight. They are either on balance or off. Keep the legs apart, get down over your sheep, becoming almost part of it, and handle it with body co-ordination resembling a graceful slow waltzing movement. It will take young shearers years to get this but it is the goal to aim at.

It takes ten years completely to master shearing, five years to learn to hold a sheep, and another five years to perfect all aspects of the job.

Nothing will tend to eliminate second cuts more than correct position pattern shearing.

3 MINIMUM NUMBER OF BLOWS

A minimum of blows applied on the sheep should be aimed at. No definite figure can be set down, as blows differ, breeds differ, and shearing differs. On good average-shearing crossbred sheep anything from fifty to sixty blows is satisfactory. There have been some experienced shearers who take over ninety blows for every sheep. This means that to shear the same tally their hand must move twice as fast, which is not a good practice. In my travels around this country I have met those who boasted of shearing every sheep in thirty-six to forty blows, or who know someone who did. All I can say is that it is a bit ridiculous. One must be moderate, and to try and go to the extreme to save blows, at the cost of physical effort and difficulty, is also an error. There are places on a sheep, such as the crutch, where perhaps four blows are better than two, and also safer for the teats. Remember that if in odd places you can do two blows more easily than one then certainly do two. What is really meant by seeking a minimum of blows is that you should fill the comb so that you are using it in most places as a full comb and not as half a comb.

4 Use of the Left Hand

The left hand must be used as much as possible. It is one of the most important parts of the body in shearing, having the job of preparing the way in front of the handpiece. The secret, then, is to hold the sheep as much as possible with the legs, to leave the left hand free to do this work. The reason for the left hand's not being used enough is that shearers are afraid of cutting the hand. A handpiece is a dangerous tool, and can inflict a bad injury, but my advice is to train oneself to have confidence with the hand in close proximity to the handpiece. To demonstrate this point, it is necessary only to watch a 'gun' shearer in action to see the way in which the left hand is co-ordinated with the handpiece, and in several places on the sheep is really close to it.

5 Never Fight a Sheep

One must never fight a sheep in the course of shearing. Yet how hard it is for a shearer to control his feelings! No one knows better than I how a bad run can tend to get a man down—a run of sticky sheep, or gear not cutting the best, or perhaps the back aching a bit, and a lot of other worries (too numerous to mention), and suddenly a big wether kicks and his sharp hoof claw rakes you down the middle, or the handpiece flies and breaks a good comb. A man needs a lot of grace to control himself. And yet the worst thing a shearer can do is to lose his temper, fight the sheep, perhaps bash it over the head with the handpiece (not good for the sheep or the gear), and work himself into an excited frenzy.

Remember that there is a fresh sheep waiting in the pen, and it is like a wrestler taking on a fresh man every round for a two hundred round fight. *Don't fight sheep.*

Blade Shearing

It is important to realise how much blade shearing is still done all over the world. In the high mountain country of the South Island of New Zealand the use of blades is still preferred by most woolgrowers. Compared with machine-shorn sheep the wool left on the sheep after they have been shorn by blades offers a much greater protection against the sudden cold snaps or storms that can come up at almost any time of the year in those regions. All these high country sheep are in the fine-wool class, predominately Merinos with a limited number of Half-breds and Corriedales. (Blade shearers much prefer fine to coarse wool, as the former is easier and softer to cut.)

I have not seen or heard of any blade shearing in Australia but most of the sheep in South Africa are shorn in this manner. The native shearers there are not machine minded; they prefer blades and as most of the sheep are run on high country the growers see no reason to change. (The natives, incidentally, shear sheep at 1d. each.)

In many parts of Britain, particularly in Wales, Northern England, Scotland and Ireland, it has always been the practice to use blades. There are many small flocks in these regions which are usually shorn in the open and blade shearing persists largely because of the difficulty and cost of setting up machines in the fields for limited numbers of sheep. Again, many British flockmasters like to see some wool left on the sheep. On my two visits to Britain, however, I noticed that machines were becoming increasingly popular, and if the present trend continues it will not be many years before machine shearing will be the accepted British method.

In Europe there is still much blade shearing done, but again largely because of small flocks. Machine shearing is preferred wherever possible. In North and South America blades are used in isolated areas though machines have been widely adopted in the last twenty years.

Generally speaking, there is still a great deal of blade shearing done throughout the world, and though it seems unlikely this method will ever be finally superseded, there is nothing to prevent the increasing use of machines in all countries. Many will agree that this is 'a good thing', though of course there will always be some, particularly high-country men, who will shake their heads and even dispute the notion. But there is no point in not accepting the fact that we live in a machine age and with shearing, as with many other occupations, it is inevitable that more and more people will become convinced of the advantages given by machines.

Wherever machines have finally taken over from blades, however, it will always be necessary to treasure the memory of the blade shearers as one does other aspects of pioneer life. Some great stories can be told by blade men, and a good blade shearer has always been pretty to watch.

Blades *versus* Machines

This has been a much-debated subject, and can be likened to the old argument of the horse *versus* the tractor. The most marked difference between these two types of shearing is in the handling of the sheep. A good blade shearer works around his sheep, and can shear it in almost any position on the board. If it should struggle or kick one way or the other this does not worry a blade man.

It is not so with machines. The machine will not come off the wall and follow you round the board: the sheep must be in the correct position for machine shearing. To be in this position all the time while being shorn it has to turn almost completely round (or nearly one revolution) in the shearer's legs. This is why machine shearing is harder, or takes longer to learn, than blade shearing, but when it is mastered it is probably much easier. It is, no doubt, one of the reasons why blades surpassed machines when machines first came in, even though the gear was rough. If a sheep kicks out of position in machine shearing you have no option but to stop and haul it back, merely because the dropper will not reach its objective.

With blades the relation of catching pen door to the board or porthole is not so important as with machines. If the catching door is behind a machine shearer, he has to drag every sheep forward up to the machine (you can take a sheep back but not forward) as there is only one right place on the board to shear with machines. As to the gear, I think the machines have the advantage. Grinding does not take long with the machine, but with blades, hand turning the old grindstone for your mate, and then vice versa, is a much longer and more tedious job. Machine men only stop a few times a run to change cutters, compared with blade men giving their shears several wipes in a run.

In the noise of the shed there is a vast difference—blades silent and machines noisy. It is amusing to hear different opinions. The blade men cannot stand the noise of the machines, saying it gets on their nerves, while the machine men cannot stand the silence, and say it is boring and lacks action. They claim the sound of the machines going makes everyone go to keep up with them. It all depends on what one is brought up to.

Relative quality of workmanship between blades and machines is a debatable point. There are good and bad jobs done in both classes of shearing. To a flockmaster used to machines, blade shearing leaves too much wool on—the sheep not smooth enough. To one used to blades, machine-shorn sheep are much too bare—nothing left for protection against the weather. With either blades or machines, however, a good job is easy to look at and a rough job best put as far from view as possible.

Now as to which is the faster. In this there is no real comparison, for machines are much faster. One knows of big tallies with blades, a record of over three hundred in Australia and around that figure in New Zealand, but it must be taken into account that sheep in those early blade days had not the wool or trimmings our present-day sheep have, and it is the trimmings on legs, head, and crutch that slow blades up. In the old days, there were a lot of bush burns, with the result that bare bellies and bare points were much more frequent than today. We have spent years on the breeding of sheep and on obtaining improved wool. The present-day Romney, with wool to the toe and even on the 'top lip', is much different from the clear open-pointed Romney of fifty years ago.

However, it is not for me or for any man to decry the abilities of either machine or blade shearers, as whether he uses one method or the other on a sheep, a shearer is working hard and doing a man's job, and has his heart in the right place. But I firmly believe from what I have heard and read of the conditions that early blade shearers had to put up with, that they were more iron men than we are today.

Thirty years ago an aged English farmer came out to New Zealand and took up a small farm. He milked cows the first two years, and then, because of lack of labour, he went into sheep, putting on 250 ewes. Shearing was left until late in the season. There was a small power plant on the place but he had never seen one—only having seen blade shearing done in England. A friend of mine was asked to go out and do him the favour of shearing his flock. He duly arrived. The old chap said, 'How do you shear—by the hour, the day, or the week?' When told by the day, he asked, 'How much a day?' My friend replied, 'About ten pounds.' 'My word,' said the old chap, 'it's going to be a costly job—four or five days.' My friend said, 'Wait till the job's finished.' He then hopped into the pen and peeled off a couple of sheep. The old chap stood transfixed with amazement. He then went outside and shouted for his wife and daughter to come and see the miracle man. My friend was a bit of a wag, and while the old man was away, he shore the topknot wool off the sheep's head and patted it back on the poll of the head. Then, when the three had come back and were watching him shear, he came to the head, gave a mighty puff of his cheeks, and the wool flew off. 'He's blowing it off now,' said the startled old man.

The Snow Comb

In recent years the increased use of machines in many of the cold high-country woolgrowing areas of the world where sheep require some protection after shearing has led to the development, by myself and others, of a high-running snow comb. The problem was how to design a type of comb which would allow the wool to be cut further away from the skin where, as every shearer knows, penetration becomes more difficult. The present snow comb is made with large sledges on the base of the outside and every other inside tooth, so that in use it leaves a good ¼ in. to ½ in. covering of wool on the sheep (equal to three weeks growth). As regards penetration, four of the thirteen teeth on the standard wide comb have been removed to make this easier. (Again, as all shearers know, on hard shearing thin or worn-down combs go much better.)

On all fine-wool breeds the snow comb has done, and will continue to do, a good job. Sheep shorn with this comb have stood up to some severe storms in New Zealand immediately after shearing and only very minor losses have been reported, in most cases no losses at all.

When it was first introduced the comb came in for a certain amount of criticism both from shearers who found it much harder to use and from woolgrowers who still favoured blades. In the first two years the pointer swung for and against and then for again. It should be realised, however, that the shearers were new to its use, in many cases shearing reluctantly with it. The combs themselves were all new and no shearer likes a new comb; it does not comb or enter like a worn one.

There are now many shearers expert in the use of the snow comb and many farmers who favour snow-comb shearing. It has been of advantage to both, giving the shearer a much longer season as well as allowing the farmer to shear at any time of the year without undue risk. Stud flocks especially are suited to snow-comb shearing.

To anyone using this comb for the first time I would recommend that he persevere with it for two or three days as it is essential to use a slightly different angle to the sheep to give the desired finish. Blows should be made more with a brushing style so as to leave wool on, rather than with a poking style to take wool off.

It will be found that the comb will be much easier to use when it has been ground down to where the scallop on the comb teeth has disappeared. In hard or 'sticky' shearing to take any shoulder off the front of the teeth will also greatly assist penetration. New combs are now being made with no scallop and no shoulder.

Once mastered the snow comb does a good job and answers well the purpose for which it was designed. However, where it is not required that wool be left on the sheep the standard comb is the one to use.

For the carpet wools in Britain, i.e. Blackface, Swaledale and other coarse-wool crosses, the snow comb has too wide a gap between teeth and one of the thicker standard combs will make a better job. With open coarse carpet wools penetration is not such a problem.

Almost any cutter can be used with the snow comb, but one with wide-base teeth runs much better as there is more metal to cover the extra gap between comb teeth. The snow comb is used with much brighter points than standard gear. The comb's points should not be rounded, since in combing off the sheep there is little danger of cutting the skin. ('Lead' is adjusted as on standard gear, see page 92.)

CHAPTER TWO

The Sheepowner: Shearers' Diet and Clothing: Runs

Sheepowners' Obligations

SHEARERS AND THEIR ATTRIBUTES are usually discussed much more than sheepowners, their woolsheds and general facilities for shearing. While acknowledging the sterling qualities of most sheepowners, it is imperative that I set out the obligations and requirements of a sheepowner. The sheepowner may be a man with a small flock or one with a large holding, and with such a vast difference between the two, I cannot deal with this subject completely. But the main basic points apply to all. The responsibility of shearing cuts two ways, and the boss must be prepared to meet the shearer in reasonable requirements and give him the co-operation and backing he deserves.

One of the first requirements in a woolshed is a decent grinder. How can a man shear if he can't get his gear to cut? —and he cannot get his gear to cut without a proper grinder. If you walk into a shed and see a bad grinder, and then go outside, you will see that bad grinder reflected in the quality of the finished work. (See section on grinder installation, page 82.) The next point is decent emery paper. There are too many worn-out papers used in grinding. In many cases, especially in smaller flocks, the farmer is reluctant to change or renew emery paper, and yet the cost is small. Farmers should not stint grinding paper and they will get a better and happier job from their shearers (see page 84).

The next point, although it only applies to sheep in heavier rainfall areas, relates to 'daggy' sheep. Sheep should be dagged before they enter the woolshed, first for the good of the farmer, and then for the shearer. Any farmer knows that in trying to pull dags off a fleece on the wool table a lot of valuable wool is wasted, and at the end of the day, a great pile of wool and dirty dags presents itself in the corner. This fault is avoided if sheep are dagged prior to shearing.

Then what about the shearer? He has his comb going nicely and in half a dozen daggy sheep the edge is gone, and probably the comb is score-marked. No wonder a shearer loses his patience on daggy sheep. The effect of not having the required co-operation is that he naturally does not put the same care into his work. Tramping on hard dags with moccasins can give a shearer a bad stone bruise.

And now the woolshed. A shearer requires his catching pen and door in the right place in relation to his machine. (See Woolshed Planning, page 134.) How many sheds are wrongly laid out, it apparently being considered that anything will do for the shearer! He requires light, either by a window on the wall or a skylight in the roof in front of him (Cooper-style louvres are an ideal window for a wall). He requires as much fresh air as possible and yet no draught. For this reason internal boards in the high part of the shed and good ventilators are preferred. He requires a board of the right width so that he doesn't have to drag his sheep too far, and a catching pen of the right size so that he doesn't have to play 'rounders' with the last sheep. These are his many requirements which are simple to provide.

Shed hygiene is another important factor, and dirty smelly skins, old manure sacks, and other litter should be cleaned up before shearing. Scrubbing the shearing board is another essential—one which takes very little time yet makes conditions much more pleasant for the shearer.

Shearers' quarters should be up to the mark. In the past there have been some bad failures in this respect. It is essential for the shearer to have a hot bath or shower every night, and a good comfortable bed to rest his back and tired muscles.

Whether it be contract gang shearing or individual shearing, 'smokos' and meals must be on time for the shearer. If he has to stand around waiting for 'smoko' or meals he loses the value of the rest period, and loses the time necessary to digest or even partly digest his food before bending again.

On a cold day the shearer requires the sheep packed up to warm them. This makes a big difference to the shearing, especially on sticky sheep. On small fat-lamb flocks, sheep brought to yards in small mobs straight off the paddock puffing and gasping make the shearing difficult: I have seen many sheep die of gasping this way. (*Note for shearers:* If a sheep is gasping, let it back in the pen rather than try to finish it and lose it on the board.) There will be many readers who do not understand this, but I assure you in all fat-lamb areas, gasping full sheep are a problem. They should be in yards for three or four hours at least before shearing. Lambs, of course, are better shearing right off the paddock. Some big mobs are in yards for two or three days before shearing. If possible they should be let out for a bite.

A shepherd can practically work in with his shearers in the points mentioned here. These few remarks on shepherding will not apply to many areas in Australia and South Africa, and in Central Otago of New Zealand, but are put in for the benefit of those to whom they apply. It is extremely hard to cover sheep-farming as a whole, owing to the difference in each country and district, but I write with the hope there is something for all.

Some readers may think I am being too particular in writing on such points as these. My answer is: 'Shear for a week or a month and you will find it all true, and essential.'

Shearers' Diet

It is rather hard to set out a diet. However, I will give first the best and most essential basic foods. These are plenty of good meat, plenty of cooked greens to keep the blood right, and any of the common vegetables. For the lighter side, cold or cool sweets are the best. Plum duff is not good to shear on. Stewed fruit or milk puddings are more on the right lines. A shearer prefers cold stewed fruit for breakfast rather than hot porridge, i.e. if he has done an early morning run. A shearer should not start an early morning run without a cup of tea and something to eat. It is essential to have something in the stomach before shearing or a shearer soon finds himself getting weak. Bending over, a shearer loses his appetite—yet he must eat, and for this reason food should be made as attractive as possible.

When actually shearing—that is, in the runs—a shearer should not drink too much. With so much sweating, he works up a tremendous thirst, and a shearer could drink twice the amount of liquid as a normal person. Although it is a great temptation, he must guard against drinking while shearing and train himself to take only one or two swallows every half hour or so: even if he rinses his mouth, he need not swallow liquid. Cold drink on a hot stomach will soon affect any man. At the end of his run he can satisfy his thirst, but still moderately, and at the end of the day, to his heart's content (don't get me wrong, as I am a teetotaller). I would start a real debate if I were to suggest the right liquid for a shearer to shear on, but my own personal preference is glucose and pure lemon juice (or lemon extract mixed with plenty of glucose). Just as tasty as all your fancy drinks, and more beneficial.

Shearers should take a blood tonic, as with moving around into different conditions, different water, etc., the blood tends to get out of order. A good blood tonic goes a long way in eliminating the danger of boils and grease boils.

Shearers' Dress

How many grand shearers tend to miss out on this point. The first essential is woollen tweed trousers, which will absorb the grease. Thin cotton material lets grease through and is dangerous to the shearer. If sheep's grease, being slightly poisonous, enters the pores of the skin, grease boils and skin troubles arise. It is necessary that trousers fit properly: too tight trousers allow no room for bending movements, and too loose and baggy trousers interfere with the sheep. Personally, being of stout proportions, I usually have my trousers tailor-made to my liking, which costs a little extra for each pair and yet makes such a difference to the comfort of shearing.

Next is the singlet—and this is a problem. One needs a woollen singlet to absorb the sweat and yet in hot weather it is almost unbearable. We need a sleeveless singlet designed with a good long woollen back and a light cool front with big arm holes for free movement. To keep a good back and good health wool is essential on the back, and a woollen sleeveless singlet should be worn whenever possible.

Should one wear braces or belt? This, I should say, all depends on the physical build of the shearer. I was always too big round the chest and shoulders, and braces chafed and impaired my shoulder action. But as braces do not collect the sweat round the back as a belt does, I think they are the best if you can wear them. If not, I found the best type of belt—and one which worked excellently—was a good wide piece of car tube which gave as one bent and did not collect sweat.

Sack or skin moccasins are the correct footwear. I would put sack ahead of skin: they do not last as long but it is better to have a good clean pair every few days than an old greasy pair. The next section describes the correct way of making moccasins, and all shearers should use this type of footwear in preference to sandshoes, slippers, shoes, or bare feet, all of which I have seen used.

'Bowyangs' are essential to shearing, more so than plus-fours are to golf. Straps are better than twine and will not cut off blood circulation. They are also comfortable and easy to fix. Put around the trouser legs under the knees, they give room for bending. Without bowyangs, and with sheep working on the legs all the time, trousers would soon work down. Many times I have shorn behind a cobber and wondered when this was going to happen. Woollen socks folded over trouser ends leave a neat working appearance. If a shearer finds it very hot, to remove socks is a help. A small clean towel for a sweat rag hanging from a nail by the pen door will allow the shearer to wipe the sweat off without losing time. A woollen jersey to put on during spell breaks is essential to avoid risks of a chill.

Making Moccasins

The materials required to make two pairs of moccasins are: one good sack, five lengths of twine, a packing needle, and blade shears.

Laying the sack on the floor, cut it as in *Figure* 1A. Use both thicknesses of sack for each foot. Now take the cut piece of sack (double thickness) and, placing the foot on it, with heel one inch in from the back of sack, trim as *Figure* 1B at about three inches off the foot all round. The inside layer of the sack is next cut closer in to the toes, approximately one and a half inches out (see dotted line in *Figure* 1B). This allows the bottom layer to fold over the top layer at the toes, thereby making a neat toe to the moccasin, which otherwise would be too bulky.

The straight back of the piece of sacking is next folded in half and stitched up to within three-quarters of an inch of the bottom as in *Figure* 1C.

Now, placing the foot in the sack with the heel hard back into the stitched part, gather the sack around the toes in four tucks, as follows: First place a full length of twine in the needle, using it double. Now, as in *Figure* 1D, fold the bottom layer of the sack over the top, and, starting on the right side, fold into a neat tuck and push the needle straight through the formed tuck. From here make tuck 2, tuck 3, and tuck 4, going round the toes, taking up the slack or loose sack.

When the four tucks are threaded, pull the twine through half-way, and then pull it up tight and tie in a knot, as in *Figure* 1E. Now thread the twine across the top of the foot through the side of moccasin and back again through the other side. Cut the twine close up to the needle, thread the other half, and put it through the opposite sides to form cross laces, and the completed moccasin is as in *Figure* 1F.

To put on, the strings are pulled up tight, tied at front of the ankle and around the ankle.

It is a waste of time making moccasins of poor sacks or of single thickness as they will wear through in a few hours. From a good sack, however, and using double thickness, a pair of moccasins will last four or five days.

Shearers' Runs

In different countries and districts, standard hours and runs of shearing are not the same.

In New Zealand the shearers' award sets out a nine-hour day.

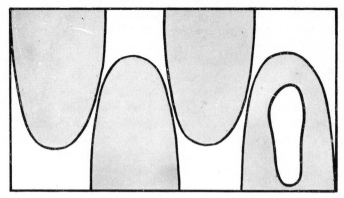

(A) *Cutting the pattern from the double thickness of sacking.*

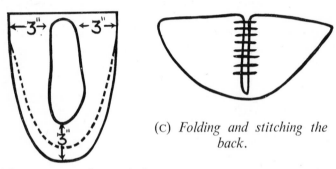

(B) *Trimming the inside layer.*

(C) *Folding and stitching the back.*

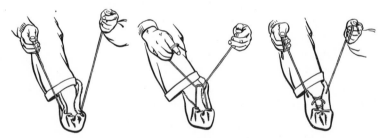

(D) *Threading the tucks.* (E) *Tightening the tucks.* (F) *Cross-lacing.*

FIG. 1. *Making moccasins.*

In Australia the award sets out an eight-hour day. While a nine-hour day is mostly worked in the North Island of New Zealand most of the South Island would work only an eight-hour day. Nine-hour days are by far the most popular (and really necessary) in high rainfall areas. Shearers can sometimes sit around for weeks waiting for sheep to dry, resulting in shearing getting so far behind schedule that when sheep are dry it is imperative that nine hours be worked.

Having tried both days, as a shearer I would still prefer a nine-hour day. It gives a man a great start for the day, when he walks over to breakfast with a good number of sheep behind him. He also enjoys this meal much more than getting out of bed for it. I also believe that a two-hour run is too long to shear without a break.

As I have been asked so many times what I consider is the best way for a shearer to work his hours, I set out the following nine-hour day, realising of course that there are many other good ways of working this, and that an eight-hour day is the one most often worked in the world shearing activities:

START	RUN	STOP	BREAK
5.30 a.m.	$1\frac{1}{2}$ hr.	7 a.m.	1 hr.
8.0 a.m.	$1\frac{3}{4}$ hr.	9.45 a.m.	$\frac{1}{2}$ hr.
10.15 a.m.	$1\frac{3}{4}$ hr.	12 noon	50 min.
12.50 p.m.	$1\frac{3}{4}$ hr.	2.35 p.m.	$\frac{1}{2}$ hr.
3.5 p.m.	$1\frac{3}{4}$ hr.	4.50 p.m.	10 min.
5.0 p.m.	$\frac{1}{2}$ hr.	5.30 p.m.	—

Total Shearing Time 9 hours   *Total Spells* 3 hours

It is in the last half-hour that the shearers sprint for home and really show the board what they can do. With this short pleasant run a man finishes the day feeling satisfied with his job and with his tally.

I trust readers and official bodies in other countries who do not agree with a nine-hour day will view in a practical light what I have written here. No concrete statements as to suitable runs can be made without being fully acquainted with climatic conditions, and with the special problems governing shearing in each country and in its various districts.

CHAPTER THREE

Shearing the Sheep

NOTE: *For the photographs illustrating this section a Romney ewe carrying eight months' wool, and clipping 8 lb., was used, thereby enabling the various positions to be shown more clearly than would have been possible with a sheep carrying a full year's fleece.*

---

1. The Catch

THE FIRST PART of shearing is to catch your sheep and get it on to the board in the correct starting position. There are several ways of catching sheep:

1 Turning the head, dropping the sheep on its side, and then dragging it out by the front legs.

2 Legging the sheep out on the board. The sheep is caught by the hind leg, and, with the other hand on the back of the sheep, it is dragged out backwards.

3 Diving into a pen and grabbing sheep anywhere, just hoping for the best.

4 The modern and correct method (*Fig. 2*). In this preferred method the sheep is caught under the throat by the man standing directly behind it with one foot on either side of the sheep's hind feet. With a smart upward lift of the arm the sheep's fore end is raised off the floor, and the man walks back quickly keeping the sheep held out at a forty-five degree angle on its hind legs. In trying to sit down, it will move its hind legs and walk out of its own accord. One naturally needs practice at this. Balance and surprise play a big part. Choose a sheep with its head held in a natural position in the pen:

the sheep with its head down is taken by surprise when the opportunity arises. There are a few in every hundred sheep who will slip right down and one has to drag them out. However, nothing is lost as this procedure is still easier than other methods.

It will be said that on big sheep only strong men can do this. This has been proved false, as I have taught and seen many average men catch big sheep in this manner. If sheep are a bit

FIG. 2. *The catch*

stubborn on rising off the grating, press the fingers of the hand
up under the jaw, and the sheep should then respond.

Advantages of this style of catching over others are:

1 It is easier on the man.

2 It saves time.

3 The sheep arrives in position without struggling or upset.

4 For the sheepowner it is much better for the wool and the
sheep, especially at crutching time, when ewes are in lamb.

2. The Belly

NOTE: *In this text the sheep's right legs and right side are
referred to as the 'near' and the sheep's left side and legs
as the 'off'. The 'near' side, in fact, is the side of the sheep
nearest the machine when the sheep is in the starting position for shearing.*

There are several different ways of shearing the belly, and
shearers differ perhaps more concerning this part of the sheep
than with any other. Before machines came in the blade style
was to sit the sheep upright, put the front leg of the sheep under
the shearer's arm, go down the near side out the flank, turn
and come back up to the brisket, shear this in upward short
blows, and then go round and round the belly. This takes a
lot of blows, and costs a lot of effort, but is still practised by
many. From this style, shearers started to 'spear' the belly off,
usually by laying the sheep almost flat on its back. This style
has failed, as the shearer is in a wrong position to do the
crutch and go down the off leg. From there the faster men
developed spearing the belly off with the sheep in a more upright
position. Wide and varied are the many styles of doing this
by the many hundreds of shearers I have seen in my shearing
experience.

The style of shearing the belly set out below is that which
I personally practise, and is easy and simple to learn.

The first point is that the sheep must be in the right starting
position; that is, directly opposite the handpiece when it is
lying on the board in a natural forward position connected to
the down tube.

The second point is the lie of the sheep (*Fig. 3*): this is very important. You will never take the belly off in this manner unless you start with the sheep lying into the porthole wall, and keep it on that lie for the whole of the operation. The starting point at the brisket is six inches lower than if the sheep is sitting straight up, thereby giving a much easier start. The sheep will not struggle as much, as no sheep will sit or lie well upright on the point of its tail. The trip-cord of the machine is put to within eighteen inches of the floor, and to switch on, the trip-cord is pulled and the handpiece is picked

FIG. 3. *The start of the belly*

FIG. 4. *The belly (mid-way)*

up in the same downward motion of the 'right hand'. To do this requires only confidence, and even learners should practise this method of switching on.

The left forearm brushes the wool on the brisket up (parting it for an open start) as the left hand takes the near front leg (*Fig. 3*). Note that the shearer is more on the side of the sheep, with feet well spaced, rather than with feet close together taking the two legs and bending right over it.

The first blow starts at the top of the brisket, clearing the near side of the brisket, and finishes down in a full comb to

the near flank or just inside it. As the belly lies in to the porthole this blow seems slightly to curve out and then in to the
flank. This is because of the lie of the sheep. On Merinos the
sheep has to be stretched up by a lifting pressure by the left
hand on the near front leg. The belly wool is not broken out
by this blow.

The next blow is from the off side of the brisket down to the
opposite flank, and if possible the belly wool is broken out.
If this is found too difficult to break out, then leave it for the
third blow. On all sheep, as the comb goes from the brisket

FIG. 5. *Completing the belly*

to the paunch of the sheep, the comb will dig in if the sheep is not again stretched up and slightly backward to take the wrinkle out. This second blow is the most difficult to master in the whole of shearing, and is probably the only reason why 'spearing' the belly off is found difficult to those who first try it.

The next blow is down and across in the same direction. As the fourth blow is going in (*Fig. 4*) the left hand folds the near front leg and places it through and behind the shearer's right knee.

NOTE: It is important that this leg must be folded away behind the knee (*Fig. 4*). The sheep is now held securely with the legs in a perfect position for the crutch and the off leg, and the left hand is free to go down on the belly where the fingers stretch the skin up to finish the belly in downward blows—and not in blows running around the belly (*Fig. 5*). The belly, finished easily and quickly with a minimum of blows, is taken off on the near side of the sheep.

On a wether the third blow is put in just as on a ewe, but make sure you are at least one inch on the off side of the pizzle. The fourth blow is not put in in the same way, but goes only part of the way. Then another blow is made down on the near side and the pizzle is found by the handpiece coming around to it on the near side, cleaning it from the side instead of straight down across it. One seldom ever marks or cuts a pizzle this way. I know of many 'gun' shearers who go straight down across it, and while perhaps this may be all right for odd 'guns' it is not a good practice for the majority of shearers or learners, as it results in too many cut pizzles.

In shearing lambs, of course style differs slightly but is the same in principle, and any shearer following the above methods will soon adapt himself to lamb shearing, with perhaps special blows or holds here and there.

3. The Crutch

The near front leg is still behind the shearer's right knee (sheep lying in to the machine). Now, from the finish of the belly run out the near hind leg. It is not necessary to shear the near flank at this stage as it can be got at easier and better by coming down across it on the last side. After running out the leg bring a blow from the end of the near leg or hock around into the centre of the crutch, and follow with another one to the centre. While these blows are being done, the left hand is in the near flank to straighten the joint if necessary. After these blows are completed, on ewe lambs, hoggets, and two-tooths the fingers of the left hand cover the teats to avoid cutting, and the third blow is run out inside the off leg. The left hand now moves to the off flank and the fourth blow goes out on the off leg to complete the crutch (*Fig. 6*).

FIG. 6. *Completing the crutch*

The crutch, of course, could be done in two blows right round, but it will be found that four are easier (especially for learners). It is just as quick to do this, and also much safer for teats. On bare-pointed sheep or wet ewes, with large udders and a reasonably bare crutch, it will be found easier and better to run blows straight down instead of around.

NOTE: Always take out a good clean crutch, as it makes for a fast off leg, and fast finish on the last side, and the wool inside the legs is more easily shorn in the crutch than in either of these other places.

On ewe lambs and hoggets it is not necessary completely to clear the wool around the teats. For young shearers especially, it is better to leave a bit of wool here than to run the risk of cutting a teat.

4. The First Hind Leg

The sheep is almost in the same position as when started, leaning in to the porthole, with its front leg still behind the shearer's right leg. On an average woolly-pointed sheep the first blow of the off leg is out the top of the leg to part the wool over for the next blow, which turns from the first one at the end of the leg and comes back into the flank. This way the comb enters nicely on the off leg and it takes little time to run out. The left hand (*Fig. 6*) is in the flank to straighten the leg if necessary. In this second blow, that comes down to the flank, the better shearers will just clear the flank. Younger shearers will have a slight difficulty in doing this. As long as you get to the flank, you are all right, for when you come to the long blow, with the sheep laid out correctly, you can get the flank quite easily at that time—at the start of the long blow. Time can be wasted trying to clear this flank, and sheep can easily be badly cut in this place. However, the ideal is just to clear the flank with this second blow.

The third blow on the off leg comes from the end of the leg or hock down to just below the flank, and then on straight down diagonally, almost to the backbone. This is important (*Fig. 7*). The comb is kept on the firm part of the leg and does not run up into the paunch of the sheep. The line for the long blow start is a straight line with a clean cut, devoid of second cuts. The next blows go from the hock in the same downward direction (the leg shorn to its natural pattern), to the near side of the backbone.

IMPORTANT NOTE: At the end of these three or four downward blows the right hand and wrist must tip out to keep the comb on the sheep (*Fig. 7*). On the last of these blows down to the tail the shearer moves back with his feet, allowing the sheep to roll into the machine and thereby bringing its tail and back

FIG. 7. *The start of the off leg. 'A' marks belly wool*

FIG. 8. *The finish of the off leg*

higher off the floor (compare *Figs. 7* and *8*). Now make a
cleaning blow across under the tail, then three blows starting
from the tail and travelling up the back in line with the backbone to the loin of the sheep. (The reason for these blows going
well up the back is to give a short, easier long blow.) The object
is to clear the top of the tail and, if the sheep is lying well, to
endeavour to clear the far side of the tail. On extra-long-wool
sheep that have not been fully crutched this back wool will
wrap around the back of the sheep and create a difficulty at
the finish of the sheep, i.e. when finishing the last hind leg. On

this type of sheep it is essential to shear well over the tail when doing the off leg. On sheep like the Scottish Blackface, with long carpet wool, the shearer has to shear well over to the far hip.

If the shearer leans forward these blows can be easily applied. As the sheep is lying well down on the floor, the shearer's right knee must be kept under the jaw of the sheep, locking its head

FIG. 9. *The completed off leg. The wool has been broken away specially to show the leg pattern*

back against the shoulder to keep it from struggling or getting up. (See *Figs. 8* and *9*.) Note also in *Figure 8* the left hand on the side of the sheep folding the wool back so that the shearer can see these blows going in, allowing him to keep the comb on the sheep.

This technique is entirely new and differs from the usual method in which blows travel from the hock up into the paunch of the sheep, and where the left hand lifts the hock as the shearer puts short blows in over the tail. By the older method the shearing of this leg is much harder and slower if anything than by the new method, and also the shearer still has a long awkward long blow, almost from the tail to the head, and tends to make more second cuts than with the new style.

On Merino sheep that show wrinkles over the tail it is easier to lift the hock to stretch the wrinkles, but the shearer should still endeavour to get well up the back, as the new style of shorter, higher long blow suits Merinos. Also, on very thin crossbred sheep it will perhaps be found better to lift the hock, for on thin sheep it is essential to clear the far hip-bone. However, this new style should be adopted wherever possible as shearers will find it an advantage.

The way a sheep looks when this new style is properly done is shown in *Figure 9*. Here the shearing pattern is shown completed with a full comb and without second cut. The fleece had to be broken and parted from the uncut wool to show this leg pattern, but, as will be seen in *Figure 8*, the fleece is not normally broken or parted.

5. The Topknot and Shift to the Neck Position

From over the tail the sheep moves forward from lying down on the floor to an upright position in the shearer's legs for the neck. Experienced shearers should endeavour to do this shift with the legs, allowing the hands free to run two blows on the head, removing the topknot wool while the sheep is on the move, coming through. This is fast, good work, calling for perfect balance and years of practice to perfect, but when mastered it is a winner. Young shearers should remove the topknot wool first and then bring the sheep up, but the procedure described above is your final objective for this part of the sheep.

The topknot wool is taken off with one blow over the top of each eye back to the ears, and if on a large well-woolled head these two blows do not clear all the topknot wool, then a further blow cleans up the centre.

For learners and young shearers a clearing blow can be put in on the far cheek underneath the eye, back towards the ear.

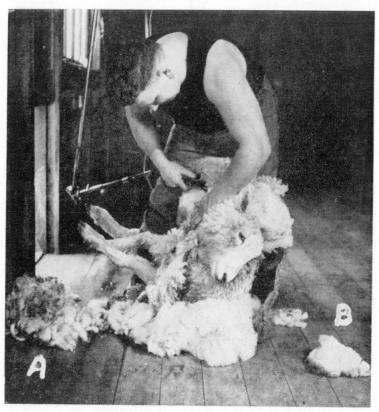

FIG. 10. *The start of the neck. 'A' marks belly wool, 'B' topknot wool*

FIG. 11. *The neck wool broken out*

In fact, on very woolly heads this is a good practice for all shearers, since it allows the left hand to rest on the side of the head when cleaning the wool on the cheek (see *Fig. 11*).

Note in *Figure 10* that the shearer's left leg is stepped well through the backbone and that the sheep is leaning out around the left thigh and knee. This is important. The shearer's right leg is between the sheep's legs: this will give the sheep a relaxed and balanced position.

6. Neck and Head

In *Figure 10* note that the sheep's legs are turned well around facing the porthole, so that the dropper is swinging clear of the shearer's right leg, and that the left hand is stretching the wool back.

The start of the neck is on the brisket. Three short smart blows are essential to open up the neck for clean shearing and a good clean break of the neck wool. One blow is made up the neck a few inches, one round into the off shoulder, and one, as in *Figure 10*, with the comb on the side to cut the morty neck wool (the first few inches up the neck) that is always a bit cotty and is the hardest to break. These three blows are done by a good shearer in very little time, and unless you were watching for them you would probably not see them go in.

From here go straight up the neck under the throat. It will be found that the neck wool, being partly cut, is easy to break out, and then, without losing control of the head, can be broken with an outward turn of the handpiece. This calls for snappy wrist action.

To break the neck wool without letting the head go takes years of practice. For shearers who are learning or not fully experienced it will be found necessary to let the head go and break the wool out with the left hand. When the neck wool is broken, the head is again held over the knee as in *Figure 11*.

Next start from back down the neck again, and come up to the side of the cheek. The handpiece does not go back down the neck any more, but keeps up at the head, one or two short blows clearing the cheek and ear, as shown in *Figure 11*. The ear is now taken in the left hand, which then rolls the head on the left knee, and one or two blows are run straight across the poll of head to the other ear (*Fig. 12*). The head should not be turned to do the cheek, i.e. coming in from nose to ear, but rather the two short blows to do this cheek and clear the ear should run from the base of jaw out to the nose (*Fig. 11*).

It will be seen in *Figures 12* and *13* that the head is cleared and yet there is still a lot of wool on the back of the neck. This is right, as it can easily be got on the long blow. It is a mistake to shear around the back of the neck while doing the head, wasting two or three blows, when the long blow gets it in the same time. There is often a lot of ground covered twice on this part of the sheep, and the boss will only pay for the first time.

This is a fast easy neck and head, and differs from the old style, in which the head was twisted into an awkward position and the neck wool was opened up behind the ear, necessitating

FIG. 12. *The head*

many short blows to do the head. It has been proved that the older style is more strenuous and slower than the modern style. On Merino sheep that have several neck wrinkles, or on wethers with horns, the neck and head form the slowest part and take a lot longer to shear and more blows than they do on crossbred sheep. On these sheep the modern style will still be found right in principle except that more of the neck is shorn with short blows following the wrinkle folds around to the backbone. Horns are not easy to get around, but the left hand does valuable work in helping to clear these obstacles.

FIG. 13. *The first off shoulder*

FIG. 14. *The start of the long blow*

7. The Off Shoulder down to the Long Blow

After the head is cleared the blows come in from the off front leg and clear the point of the shoulder. These two or three blows do not go straight around to the backbone but travel up towards the head. Three well applied blows in an upward direction will cover as much ground as six would running around the body of the sheep. The finish of these blows must be watched, however, so that the bottom tooth is on the sheep and a reasonably straight line is left for the long blow. (Second cuts are easily made in this place.) When the top side of the shoulder is cleared the left hand takes the sheep's foreleg and stretches it up (*Fig. 13*), allowing the blows to travel in a downward direction to clear the leg and underneath the shoulder.

Now run another one or two blows from the leg straight in down the shoulder, and as these blows are being done the shearer's feet will be moving a few inches at a time, bringing the sheep quietly around to the long blow position—that is, with the sheep lying straight up and down the board. The shearer will be almost facing the porthole (*Fig. 14*). When the last blow on the shoulder is being done the leg is stretched up

FIG. 15. *The long blow*

FIG. 16. *Leg position for the long blow*
demonstrated on a shorn sheep

and turned in with the wrist, and it will then be found that the
shoulder joint is flattened out against the sheep, allowing the
comb to clean it in one downward easy blow. Note also in
Figure 13 that there is no loss of control of the sheep: the elbow
of the left arm is holding the sheep's head against the knee,
while the left hand is holding the sheep's leg. The sheep is now
ready for the long blow (*Fig. 14*).

8. The Long Blow

This is probably the most fascinating part of shearing, and yet it is one of the hardest positions to master. If you see a shearer with a good long blow he is usually a good shearer. The secret of the new style is to keep the sheep high on the left leg with the left foot tucked in under the shoulder of the sheep. This is important, for the method in which the leg and foot are in the neck of the sheep is wrong, as it will not allow the sheep to roll around. I have shown the correct leg position on a shorn sheep in *Figure 16*, while actual shearing in this position is shown in *Figure 15*. It is a hard hold to get, but when mastered is very easy and is an outstanding hold, giving an excellent long blow.

The first two blows are short ones, from flank to shoulder. Then there are usually five long ones, giving one or two over the backbone. When the first two short blows are put in the right knee comes down and slightly presses on the sheep's belly, the left hand holding the sheep high but not bent around the leg as yet (*Fig. 14*). It is essential for a good finished job and lack of 'two-cut' to go level with the backbone on the long blow.

As there is only half the width of wool to take off up on the neck as on the body of sheep, it is necessary to keep level on the next three long blows (i.e. after the two short ones), to start with a full comb on the body, and to finish up with half a comb on the neck. This will keep level with the backbone, comb the neck wrinkles more easily, and give a much better job than short blows to level up. The finish of these three blows in shearing the neck on the long blow is the only part of shearing in which a full comb is not aimed at.

When the second of these long blows right up to the neck is going in, the shearer's right leg steps out from between the sheep's hind legs and is placed over the two hind legs, thereby allowing the sheep to roll around. Now, with three blows to go on the long blow, the sheep is held by the left hand behind the ears (*Fig. 15*) and the shearer is leaning back slightly, the left leg being on an angle under the sheep (and not straight up and down) with the foot under the shoulder (*Fig. 16*). The sheep is rolled around on the leg, the sheep's brisket turning to the floor. This is where you get the advantage of the blows made up the back when you were going over the tail, for you could not roll the sheep high like this if you had to reach back to the tail.

The sheep is turned around the leg only for the last three blows of the long blow; if turned too soon neck wrinkles will be created. For the first two long blows the head is held out as in *Figure 14*; for the last three it is turned around the shearer's leg as in *Figure 15*. The last two blows of the long blow are full combs (*Fig. 15*) right up to the head. Confidence here will bring the comb up to the left hand, the last long blow finishing underneath the ear.

9. The Last Cheek and Last Shoulder

When coming off the long blow the sheep must be turned with the left leg towards the porthole. Compare *Figures 15* and *17* —the sheep has come around at least a foot. This allows the dropper to swing nicely for the last side, not banging the shearer's legs, and the last side is done facing more up the board than towards the catching pens. This also allows the right knee to come into the back of the head (*Fig. 17*). The wool on the last cheek is difficult to remove when learning, as the shearer's left hand can be in the way when gripping the sheep on the head. The first blow on the last cheek is from the ear out over the eye to the nose. As this blow is being put in, the left hand grips the wool on the lower part of the cheek, which means that this blow is applied over the top of the hand. The knees play an important part in the control of the sheep in this position.

As shown in *Figure 17*, the left hand now pushes on the poll of the head and the last cheek is shorn in one blow through the cheek being crimped down. The sheep is easily held in this way. Blows then run down, finishing inside the brisket and out the shoulder inside the front leg. There is a nerve centre on the shoulder point that when pressed will straighten the leg out (*Fig. 18*), allowing the comb to run out on the leg to clean the socks. When the inside of the shoulder is cleaned the left leg steps out behind the sheep (see leg position in *Figs. 17* and *18*), the sheep's head being held through the

knees. This allows the left hand to be free to do valuable work
on this last shoulder. For the tassels right under the front leg,
especially for young shearers, the fingers of the left hand stretch
the skin bringing the tassels around so they are easier to cut off.

10. Last or Whipping Side

There are two ways of doing the last side, and there are 'gun'
shearers in both styles. One way is to keep the left leg in between

FIG. 17. *The last cheek*

the sheep's legs, the other is to step out behind. It is contended
that by staying with the leg inside the last side is crimped,
thereby saving time. In actual fact one cannot save a full blow
by crimping, maybe half a blow at most, and unless one is tall
little time is gained by it here. It costs physical effort to sit on a
sheep or dump it up on the last side. Also, having the skin
wrinkled does not make for quite such a good job, and you are
much more likely to bootlace the sheep, i.e. cut the wrinkles off
in long thin strips of skin. However, although I have seen 'guns'
of this style who are pretty to watch, I recommend my own

FIG. 18. *The last shoulder*

FIG. 19. *The last side*

style, as set out here, as being the easiest and best for the average shearer and learner. I myself, and also many other 'gun' shearers, step out on the last side. In *Figure 18* the left leg has stepped out, and it stays there with the sheep's head between the knees until the sheep is finished (*Figs. 18* and *19*).

From the shoulder the blows go in a downward direction at about a forty-five degree angle—not straight around and not straight down, but half-way between the two. Run two or three blows from the shoulder, and then the next one, as in *Figure 19*, down right out across the flank to the hock of the hind leg. This is the longest blow in this new style of shearing. Note that, as shown in *Figures 18* and *19*, to do these blows the sheep has been slightly bowed out by a small pressure on the knees. As this long last side blow crosses the flank the backs of the fingers of the left hand brush the wool in the flank out for the comb. From there carry on, finishing each blow right out on the hock, the sheep being finished at the tail and not out on the leg.

Note in *Figure 20* the left hand coming through the body to take the trip cord (which should hang eighteen inches from the floor). The machine is then switched off and the sheep, lined up to the porthole, with its head already through the legs, will walk out while the shearer's left leg takes the first stride to the pen for the next one. This is a fast switch-off, costs no effort, and wastes no time. A lot of time and effort can be wasted in switching off and pushing sheep out the porthole.

FIG. 20. *Completion of sheep—Wool away!*

It is better for learners, when the flank is reached on this last side, to bring the sheep's head back in front of the knees, as the shearer will not lose the sheep this way. But as the shearer learns and takes less time he will be able to keep the sheep's head through all the way down. If a sheep is struggling on the last side it is better to bring the head back, as it is preferable to have this momentary pause than to lose the sheep. However, when this style is practised and mastered, a shearer will seldom have to bring the head back.

11. Socks

If the front socks are required to be taken off (this rests with the sheepowner), it is best to do it before starting the first blow on the belly, taking off the two inside front leg socks with two flicks of the handpiece. This only takes a little time, and it saves time cleaning the sock when coming down on the long blow, and also saves lifting the leg up to clean the sock on the last side. This is done by following right out to the toe, using nerve centre as already mentioned. Many sheepowners do not like socks taken off, as it puts hair in the wool. On crossbred sheep, socks are mostly hair. While socks are often left on for this reason on crossbreds, the hind leg socks should be taken off to prevent sheep from getting daggy in these places. On fine-woolled sheep there is only a small percentage of hair in the socks, and on these sheep I would say take them off. In snow areas socks seriously handicap the activity of sheep.

12. Summary of Fundamental Rules for all Young Shearers

1 Concentrate on holding the sheep correctly.

2 Study position, and shear every place in the correct position of the sheep in relation to the machine. This is important.

3 Watch style, aiming at a steady controlled forward motion with a fast recovery. By not waving the handpiece about in the air, but bringing it back near the skin, it is made to cut wool and not air.

4 This style will allow you to concentrate on filling your comb. Remember also that shearing to a set style and pattern will still further help you.

5 Try to keep on the sheep or the skin, especially with the bottom tooth of the comb. This helps to eliminate second cuts.

6 Watch physical effort, and try to do the job in a way that makes shearing every part of a sheep physically easy. Shearing one sheep should be a pleasure.

7 Young shearers should not worry about speed tally or pay cheque until they have completely mastered shearing. (The history of all our 'gun' men shows that they were kept down in the first year or two.) No learner has ever been put off the board because he has shorn sheep well, but many have lost their run through over-fast, hurried, slipshod, and rough shearing.

8 Shear with the right balance and in a relaxed flexible style —not a stiff shearing arm, but with relaxed muscles and with a supple wrist. Hold the handpiece in a light grip, which gives much better touch and control. A tight grip will stiffen the whole arm up.

9 Remember that no matter how proficient a man is or how long he has been shearing, he is never too old to learn. For that reason especially, younger shearers should not disregard the advice tendered by a clean experienced shearer.

10 Never lose your temper and fight sheep, or growl and moan most of the time. This sort of thing grows on a shearer, and it takes a lot of will-power and training to overcome it. But a man can train himself to shear all day and every day with hardly a growl, and to be completely unruffled.

Remember that quality of work is paramount and is the essential qualification of a shearer.

13. Six Finer Points for Experienced Shearers

1 *Control of the sheep.* To master this, get right down over the sheep and become part of it.

2 *A positive hand.* The shearing hand must be positive not negative in its action. It must contain 'fire' and 'finish'—like a top tennis player serving an ace or a class batsman hitting a boundary. If parts of the sheep are sticky and hard to shear then attack these places positively.

3 *A good wrist.* The wrist of the shearing arm is one of the most important parts of the body. It must be supple and flexible and able to work both ways, turning to keep the bottom tooth of the comb on the sheep and bending forwards to finish every blow on the sheep.

4 *Return action.* The same as above for young shearers—fast back and near the sheep. Watch that the handpiece comes back only to the start of the blow. It is all too easy to acquire the habit of returning the handpiece inches further than necessary.

5 *The left hand.* Experienced shearers should use the left hand with confidence. Try to control the sheep as much as possible with the legs, leaving the left hand free to work wherever possible. Practise this point; confidence will come. (Many shearers do not use their ability in this connection.)

6 *Rhythm and timing.* This is difficult to explain, but really means just what it says. A good shearer can be likened to a good dancer or skater. Every movement is a perfect co-ordination of mind, body and object—combined and working smoothly together to one set purpose. The sheep is not pulled or pushed into position, but smoothly moved and turned. The shearer does not jerk through his work, shearing the sheep in sections, belly, off leg, head, long blow, etc., but runs all parts of the animal together, one into the other, so that the sheep is shorn in an even flow as oil pours out of a can.

Some writers on shearing have likened this aspect of the work to a tree waving in the wind. I can't agree with that description. To get perfect rhythm and timing, the body must not wave about too much; on the contrary, the less it waves about the better, as motion studies will prove.

Taken together as a single quality, rhythm and timing is the most difficult to learn of all these finer points of shearing. In fact, a man usually has it or not as a natural attribute. It is the quality I look for more than any other when assessing champions, perhaps not so much in competitions over a few sheep, which are more in the nature of sprint contests, but over a nine-hour day or for a record tally it is the quality that allows a man to go better the longer he works, to shear days, weeks, seasons, wasting no effort, conserving energy, always controlled, never rattled, and invariably turning out a good job. Indeed, to see rhythm and timing fully exploited one could almost put music to shearing.

The above finer points can be applied to any style, any sheep, any conditions and any type of shearer. Correctly studied, developed and practised, the sixth point, rhythm and timing, will pay great dividends.

CHAPTER FOUR

Dealing with Difficult Sheep

Tough or Sticky Sheep

EARLY IN THE SEASON, especially if sheep have had hard treatment before shearing, some sheep become really 'tough', and the wool seems glued to the skin. I myself have shorn several thousands of this kind of sheep, and have seen my tally go down from 300 a day to 150 or 170, while having to work a lot harder to get that many. Sheep like these can be very discouraging to a shearer. However, they have to be shorn, and the notes that follow will assist this type of shearing:

1 The hotter these sheep are the better they shear—they differ amazingly when they are cold and when they are hot. All openings in the shed should be closed, and when shearing starts, the sheep should be packed up tight so that they start to sweat. The 'sheepo' must keep an eye on them for smothering, the first sign of which is sheep jumping up in the pen. Shearers will immediately notice the difference caused by this packing up, and will do a much better, faster, happier job with these sticky sheep than if they are left cold and scattered about the shed. The same packing up is also very essential with sticky lambs. With a cool draught of air coming up through the grating sheep soon get sticky on a cold day.

2 An old thin comb that is nearly worn-out should be used. Such combs should be kept for this very purpose, but make sure that the comb tips, although very fine, are still rounded off (see page 89) to avoid cutting the sheep.

3 Keep the cutter right out on the cutting edge.

4 Stop more often and clean the grease off the back of the
comb, where it will quickly cake on when shearing this type
of sheep. After every four or five sheep give the comb a
quick clean, which allows it to enter much more easily. The
comb also wants plenty of surface oil.

5 Do not shear with a stiff pushing arm and hand, but rather
with a persistent flexible hand and wrist, exerting an even
pressure. It will be found that in this way you cover the
ground more quickly and more easily than with hard wild jabs.

6 Do not take a big belly off, as it is easier to get on the last
side. When going over the tail try to clear the far hip-bone,
which is most easily done at this stage. This point also
applies to very thin sheep.

The Cobbler

This is the odd tough sheep in a mob, the one that makes the
'sheepo' unpopular when a shearer finds it in his pen. It is
amusing to hear the remarks of a shearer when he gets a too
liberal share of these cobblers. I have seen a real 'toughy' go
from No. 1 pen right down the catching pens to the bottom,
as shearers threw it over the rails when no one was looking.
However, it is a bad practice to get the habit of complaining
about the deal of sheep. My advice is to shear your pen with
a smile as the good runs of sheep will be evened up with the
bad runs and it usually works out evenly for all men on the
board. This cobbler sheep is shorn in the same way as the
sticky sheep dealt with above.

NOTE: If there is a cobbler in your pen always work it in for
your 'catch', for then no time will be lost doing it. A shearer
should be allowed to save this cobbler up for quite a time before
the end of the run, leaving it in his pen when being refilled,
especially if it's an extra special cobbler. A 'catch' of course is
the sheep a shearer has on the board when the gong goes for
time up, he then being allowed to shear or finish it. Shearers
always watch the 'ringer' when time is getting near, and get
in step, as the gong often blows on the ringer, perhaps when
he has just started a sheep, for he can still shear it and be
finished before the slower men down the board.

Sandy Sheep

Sandy sheep are a real problem, for the sand, grit and dirt are hard on the cutters, combs and handpieces. In shearing them the following points will be helpful:

1 Use plenty of oil on the comb.

2 Use either old combs or new ones (i.e. ones that can stand a bit of grinding in any case), but put your good trade combs away, for these sheep will ruin good combs.

3 Be prepared to change plenty of cutters and to keep changing combs. Do not let them get too blunt. It is imperative to change much more often on these sheep than with ordinary shearing.

4 Every half-day thoroughly clean the sand and dirt out of the working parts of the front end of the handpiece, as dirt left to become bedded in will cause excessive wear.

5 In many sandy, gritty-backed mobs, there is a percentage of sheep with clean backs. Make the best of these clean sheep by using two handpieces, having good gear set up on one handpiece to shear the clean sheep. Such sheep can quickly be detected in the catching pen by parting the wool on the backs. When the clean gear is on shear the clean sheep in the pen, and when the other sandy gear is on do all the sandy sheep. This can only be done if you have two hand-pieces ready, but is a good practice.

6 Finally, a shearer should be paid a bonus for shearing sandy sheep, even if it is only to compensate him for the extra gear he has had to use and wear out, plus the excessive wear on his handpiece.

Shearing Matted Fleeces

In wet districts when there has been excessive rain, or when sheep have had a setback during the year, fleeces become very matted, especially on the necks and bellies, where it is impossible to break the wool open by physical effort, and in many cases extremely difficult for a machine to cut it open. This condition is mostly found in crossbred sheep.

Always look for the parting in the wool if there is one, and shear in unorthodox fashion, following the natural parting. For instance quite often there is a parting straight down the centre of the belly. A matted belly is usually most easily parted straight down the middle, leaving half the belly wool on each side of the fleece. No shearer can be expected to remove the belly wool from the fleece on a matted sheep (although in ordinary shearing if required to do so he must pull the belly wool off). But this matted belly is left for the wool table, where in any case it is seldom skirted. An easy way to break a matted belly is to let the blow travel right through under the wool and to break the wool with an upward lift of the forearm. If this is found too difficult, put the comb on the side and cut it open. Now the neck —the most difficult place of all on a matted fleece. It will be found that on almost all matted sheep there is a slight parting in the wool from the brisket just in front of the shoulder around the backbone, caused by the shoulder action of the sheep when walking. Another parting will be found up the middle of the back, up the back of the neck. These are the two partings to follow on the neck of a matted sheep.

When starting the neck throw the sheep away from you and come square in to the backbone, following the shoulder line in this first parting. Then, when the back parting is reached, turn diagonally up the back of the neck. This is a bit awkward to get used to for a start, but it does simplify shearing matted necks, and saves hacking and cutting through the middle of a strong neck matt.

When shearing the body of a matted sheep it will be found that the weight of wool pulls the skin out, and it is necessary to shear with a light hand to avoid badly cutting the sheep. Sheepowners must expect these sheep to be marked a bit more than when average sheep are shorn, but a light hand will keep these skin cuts to a minimum.

Quite often a shearer's wrist will give out on him when shearing matted sheep. This can be extremely painful and may seriously retard his shearing. As soon as even a slight ache is felt, a leather wristband or a tight wrist bandage should be put on to support the wrist muscles. Embrocation should also be used. A shearer will find with these remedies he can carry on with his work and the wrist will soon be cured.

Wet Sheep

A shearer must not shear wet sheep. In doing so there is more danger to the shearer than the wool. One hears a lot about wet wool heating and spoiling. In greasy wool slight dampness is not detrimental, as the grease will absorb the moisture. It is scoured wool, with all the grease removed, that is dangerous when damp and can easily heat and spoil. I have seen a lot of slightly damp greasy wool shorn and baled, but have never known of any after-effects.

The shearer, however, may contract grease boils, rheumatics and chills through shearing wet sheep. The time gained by shearing damp sheep is often lost many times by the shearer being off work as a result of it.

When are sheep too wet to shear?

1 The most simple test, if in doubt, is to shear two or three sheep. If the shearer detects even a slight coldness coming through the trousers, then the sheep are too wet.

2 If the back of the comb does not cake up with grease but keeps clean the sheep are too wet.

3 If any dampness is detected in the fleece when on the wool - table—that is, if it gives a cold instead of a warm feeling— then the sheep are too wet.

A test sometimes used is to twist some strands of wool and check the glisten of moisture by sunlight. I have tried this, but have still seen the same glisten after a month's fine weather when sheep are sweating with the heat. For this reason I would rather test, if in doubt, by the first three methods.

It is the dampness in near the skin that is the most harmful to a shearer. A heavy dew or a light shower, putting the moisture only on the tips of wool, while the inside of the wool on the skin is warm and dry, does not seriously affect a shearer or the wool.

It has been my experience that if it has been fine for a few weeks and shearers are a bit tired, it takes very little to make sheep too wet to shear. But if shearers have been waiting around to start for a few days it takes very little time to make them dry.

But shearers for their own good and health should be careful of shearing wet or damp sheep. Wind is much better for drying sheep than sun, of course, and both together provide ideal drying conditions. But four hours of wind on the face of a hill gives better drying than twelve hours of sun in the yards.

If sheep are to be dried as quickly as possible they should be kept stirred up, and not left to lie around. Sheep will also dry out a lot overnight in the shed from the heat of their bodies and from the draught coming up through the grating. Putting sheep in a good dry plantation of trees is helpful, and where there is a patch of trees in close proximity to a shed, there should be a yard under the trees where sheep can be held for drying at night.

I have been many times asked whether there is any cure for grease boils. Preventions, of course, are better than cures. Taking a blood tonic, watching the shearing of wet sheep, and, if susceptible to this trouble, wearing an added inside covering over the calves of the legs, will help to keep a shearer free from grease boils.

When a shearer does get grease boils he should rest with the legs up. Bathing in hot water is the quickest way of getting rid of them.

Shearing Rams

Rams are a small percentage of the shearing total and yet they have to be shorn and sometimes present quite a problem. It is just as well that a shearer is paid double rates for rams, for the extra work thoroughly justifies the higher rate. I would rather shear two ewes than one ram, average to average. The last thing to do in shearing a ram is to match your strength against his or to fight him. The more quietly and easily he is handled the more quietly he will sit. Give the average ram nothing to fight against and he will not usually fight. To do this it may be necessary to shear in a rather unorthodox way, but be contented as long as you are cutting wool off, and the ram is not struggling. Also be prepared to shear old style, new style or your own special style, whichever way the ram will sit the best, especially when doing the head and neck.

Pre-lamb Shearing

A practice introduced in recent years, pre-lamb shearing simply means shearing ewes from six weeks down to about one week before lambing. While there are many points in favour of pre-lamb shearing, perhaps the main one against it is that lambing usually takes place at a cold and wet time of the year. Many flockmasters are not happy at the idea of turning out shorn sheep in such wintry conditions, though it is also a fact that an in-lamb ewe, because of its higher blood temperature, is better able to withstand cold than a ewe after lambing. To a great extent, however, the snow comb has helped solve this problem of the weather.

It has been found that pre-lamb shearing saves a great deal of shepherding work. Shorn ewes lamb better, will seek out a more sheltered place for lambing, and the young lambs are better able to find the milk quickly. In addition, with a small amount of wool to carry through the hot weather a ewe will graze more freely and, according to some sheepowners, make a better job of fattening lambs. The prevalence of cast ewes is also considerably reduced by pre-lamb shearing, though of course ewes will still get cast in the late winter months. A further point in favour of the practice is that work at shearing time is made somewhat easier, since ewes without lambs can be handled in large mobs with fewer problems as regards mustering, drafting, etc. As regards wool quality, twelve months' growth of wool shorn just before lambing is in no way inferior to summer-shorn wool and in many cases is of even better quality.

All things considered, however, I do not predict any major swing to this type of shearing. It seems more natural for a sheep to lose its wool in the summer months and the great majority of sheepowners will probably always prefer summer shearing. Pre-lamb shearing does not suit a wet climate, in fact it is of doubtful value in any district receiving more than 25 in. annual rainfall. Where rainfall is heavy, grease sets hard in winter and early spring, and even on well conditioned sheep shearing becomes very difficult. The wool does not comb well and the shearer has more or less to push it off. Fine-wool breeds, of course, always shear better than coarse-wool breeds during the winter months, and the opposite applies in summer when the grease is up.

From the shearing angle there are points both for and against pre-lamb shearing. One point in its favour is that the extra work extends the shearer's season. On the other hand, the spring is not an ideal time of year for working on the board—the days are much shorter and one would have to be lucky to strike any of the warm or hot weather that shearers are used to. The shearing itself is harder, as the wool is generally sticky, and in most cases the snow comb or a high-running comb has to be used to give the sheep some protection. Besides that, ewes heavy in lamb have to be treated carefully and are generally harder to handle. It is usual for shearers to be paid a premium above award rates for pre-lamb shearing.

POINTS FOR SHEARERS TO WATCH

1 Packing sheep up (see page 53).

2 The snow comb (see page 11).

3 Lead—use little or no lead (see page 92).

4 As you are being paid a premium, and getting a longer season, handle ewes carefully and be content with a lower tally.

POINTS FOR FARMERS

1 Make sure ewes get out to feed as soon as they are shorn. It is a bad practice to keep ewes standing about too long in yards or tally pens, especially in rough weather. A sheep that has had even half an hour's feeding will stand twice as much wet and cold as a hungry sheep.

2 Save feed in a sheltered paddock especially for newly shorn sheep and hold them there for up to a week after shearing.

3 If heavy rain comes on run sheep back into the woolshed rather than leave them out. To quote one expert: 'There have been too many losses in freshly shorn sheep when the woolshed has been empty.' If rain starts while shearing is in progress, and the tally pens are not covered, shearers should co-operate with the farmer and return sheep back across the

board, i.e. returning into the shed rather than letting them out. In most cases there is not a full day's shearing involved, only the shed to cut out, and if everyone co-operates work is not held up unduly. (For individual tallies, tally catching pens at each fill.)

Second or Double Shearing

A fast-growing practice in New Zealand is to shear sheep twice a year or three times in two years, i.e. every eight months. Where a farmer favours second shearing my preference is every eight months, shearing in September or October, May or June, January or February. These are all good shearing months in New Zealand, and on present rate of growth eight months gives wool of a reasonable long staple.

Wool users all over the world have been perplexed by the sudden swing to this type of shearing. Some of the main reasons in its favour are:

1 It saves a good deal of time and work on shepherding; it is rare to have a cast sheep; flocks are more active and more easily handled; lambing troubles are halved; and lambing percentages, especially among two-tooths, are improved.

2 The earlier hoggets can be shorn in spring the better they will grow and do through the summer months. Hoggets are commonly shorn from one to two months before the ewes, which means that next year's two-tooths have much more wool than the rest of the flock—up to fourteen months' growth at their next shearing. As they have carried this extra weight of wool through the winter, there is no comparison in the size and condition of the two sheep when they come off the shears next season. The two-tooth ewe shorn again the following autumn is better grown, is in better condition, and does a better job of her lamb.

Remembering that a young sheep grows more wool, it should be realised also that a two-tooth ewe will grow as much wool in ten months as a five or six year old in twelve months.

3 The point can be debated, but there is a strong opinion among farmers that the meat concerns do not give them the full value of the wool on fat sheep picked for the works. This results in most fat wethers in autumn and winter being shorn at least three weeks before killing. (Three weeks' growth of wool is required by butchers before they will kill sheep.) Once a farmer has practised this, he seldom if ever offers sheep to the fat picker without first shearing them. In many cases only three or four months' wool is taken off.

4 It is also a fact that sheep will grow approximately an extra pound of wool in a year if shorn twice.

5 While the price received for second-shear wool is not as high as for full wool, the actual difference is not very great when one considers the fact that with second shear a much higher proportion of the clip is sold in the top line, i.e. little skirting to be taken off the fleeces and no cotted, discoloured, seedy fleeces to be taken out. When a farmer compares the total return of wool per sheep the price margin is not large enough to influence him against second shear if he is keen on it. A point to note, however, is that farmers should handle and sort second-shear wool as carefully as full wool, and employ the same number of shed hands to do the job properly. Far too much second-shear wool is just dumped into the bale holus-bolus and once mixed up it takes a lot of sorting. Not many 'seconds' have to be taken out to make a first-class line of second-shear wool, but it calls for attention, especially on the board, with a final check across the table. (On the board a good 'fleece-o' with a broom can do all that is necessary.)

On the other hand, points against second shearing are:

1 A farmer has the expense of two shearings instead of one. However, the cost of crutching should be deducted, since the second shearing takes the place of crutching. (The price of crutching is approximately a third to a half that of shearing.)

2 By halving the twelve-months' staple the overall picture and presentation of wool could be impaired. Buyers expect cross-

bred wools to have a lengthy staple. But it is too soon for anyone to gauge the full repercussions of this aspect of the question.

Altogether, time alone will tell whether second shearing is good or bad for the wool industry. The carpet trade is now showing a lively interest in second-shear wools, and other factors involved one way or another in the dispute are the new combing techniques and blending with synthetics. We know for a fact that second-shear wool has fewer defects and that the yield is considerably higher. In New Zealand, for instance, many of the longwool breeds are today growing as much wool in eight months as they did at one time in twelve months. It would seem that, looking at it from the wool angle alone, second shearing might still come out a little on the debit side, but the many advantages in the working and return of the sheep listed above probably tilt the scale in the other direction. The debate, however, is still going on.

Points for Shearers to Watch

1 Use a half-worn comb with little lead. If new combs are used do not round teeth off as much as for full wool. (See page 91.)

2 Grease will set quicker on the back of the comb, and has to be cleaned off more often than on full wool. (The quickest way to clean grease off is to rest the handpiece against a wall and remove the grease from the back of the comb and between the teeth with a screwdriver—a quick brush and the comb is again ready for action. Much time is wasted removing grease. It should never take more than a few seconds.)

3 The wool is stronger cutting and it will be necessary to change cutters and combs more frequently.
Generally speaking, second shearing is harder on gear than shearing full wool.

4 Second-shear sheep are more active on the board and the shearer has to give more attention to controlling them.

5 Like lambs, second-shear sheep need to be kept packed up. On a cold day they will become cold and sticky much quicker than sheep carrying full wool.

POINTS FOR FARMERS

1 Sheep must be dagged before shearing. The wool does not hold together to the same extent as a full fleece and this means that dags will more easily become mixed with good wool.

2 Have enough shed hands on the job to sort the wool efficiently into two lines, firsts and seconds. (This is not a shearing where shed labour can be cut down.)

The broom plays an important part in the sorting of both lambs' wool and second-shear wool; in other words, the easiest place to sort the main bulk of wool is where it comes off the sheep. A good straw broom to each stand should be the rule for fast shearers, or one broom to two stands for slower men—properly applied, this arrangement will do a first-class job on the shearing board. And remember the maxim in wool classing: 'If in doubt throw out.'

Shearing Lambs

The shearing of lambs is becoming increasingly popular, so much so that in many areas nowadays it is the exception rather than the rule to leave them unshorn. In breeding areas lambs are shorn along with the ewes, or if this is not possible they are shorn at weaning time. In fat lamb areas it is usual to shear the remaining lambs after the first two picks have gone to the works. Shorn lambs, it has been found, fatten quicker on grass and also grow and carry better through the winter. Once a farmer adopts the practice of shearing lambs, he seldom goes back to farming woolly lambs throughout the year.

The shearing of lambs can be a difficult proposition for shearers who are not used to them, mainly because they are small, very active and hard to hold. Shearers should note the following points:

1 The small size of the lamb means that there are no blows of any length and it is not necessary to concentrate on a full comb to the same extent as on big sheep. For these reasons blows can be paced up, and a good lamb shearer works with a fast, positive, light hand and a flexible wrist. In fact, the quicker his movements the better he usually is on lambs.

He is able to do this because there is little physical effort involved—a lamb can be moved and brought into position so much more easily and quickly than a fully grown sheep.

2 There is no need to be too particular over style on lambs, as each man, in handling such a small active animal, must adopt a slightly different style to suit his build. The main thing is to be able to control the lamb so that the handpiece is at all times cutting wool.

On most lambs, however, it is better to open up the neck behind the ear. This means shearing the whole head and neck when the lamb is in this position, i.e. it cuts out shearing the last cheek as on big sheep. Lambs are hard to hold when doing the last cheek, and it is advisable to shear the whole head when and where it can best be controlled.

On lambs, swing off the long blow straight on to the last shoulder. When lambs are from five to six months old and carrying plenty of wool they shear more like hoggets and can then be treated as adult sheep, i.e. open up the neck under the throat.

Little time should be wasted on the inside (i.e. belly, crutch and off leg), not much wool being taken off. The sooner one moves up to the head the better, and the speed of fast men on the inside of lambs is often amazing. Do not try to clean too closely around teats, or overtrim under the legs. Lambs' wool is short and does not hang down. A well shorn head, a good long blow and last side, with the inside correctly blocked off, turns out a nicely finished lamb. Incidentally, many sheepowners do not like overtrimming on lambs as it makes them look light in the bone.

Catch quickly, get right down over your work, and really set about the job. Lambs will not tire a shearer and he should be able to reach a much higher consistent tally than on adult sheep. As regards tallies, sheepmen do not consider lambs a fair test for a shearing record. Special lamb records have been set, but should not be confused with official world shearing records, which must be done to union rules on adult sheep. (Many lambs are shorn when only two or three months old. This is called 'whitewashing'.)

3 Gear needs to be in good shape for lambs. Poor cutting will show up more quickly than on adult sheep, as lambs' wool is denser and stronger growing, with no lift. Because the wool is short and light it will not pull the skin out, which makes it possible to shear with no lead (i.e. cutter right up to the start of the cutting edge).

For lamb shearing my preference is for thin combs rather than new combs. When the scallop has been ground out of a comb, in other words, when it has been ground out to the tip (see page 91), it should be saved for lamb shearing, on which it can be used either until it is worn out or it breaks (usually the latter). If a new comb has to be used do not round off the points too much—a very slight rounding is all that is necessary. New combs, I have found, go quite well on good shearing lambs. Lamb combs, however, are the hardest of all to get just right, and I have often noticed a shearer's run tally go up or down by as many as ten or fifteen lambs a run when using a good lamb comb as opposed to a poor one. Shearers appreciate this fact and usually take the trouble to develop good lamb combs. Remember, the brighter the point (i.e. not so rounded) the lighter the hand—the expert lamb shearer is able to make the best use of a comb with much brighter points.

4 Wherever possible try to shear lambs straight off the paddock, and if ewes and lambs are shorn together always shear the lambs first. The shearing quality of lambs deteriorates in a very short time, and if they are left standing loose in the shed or yards, with a cool wind blowing, some hard shearing is quickly in the making. Keep lambs packed up warm, as fresh off the grass as possible, and shearers will make a much better and faster job. Weaned lambs should be kept in a handy good paddock and brought into the shed in small lots. This applies especially to store mutton breeds.

Farmers understand the value of the above points and in most cases are only too ready to co-operate with shearers by giving this practical help. As regards sorting lambs' wool, the same procedure should be followed as that recommended in the previous section on second-shear wool.

NOTE: Mention was made above of packing up sheep in the woolshed. In this connection, a point to be considered is the danger of smothering. If sheep are packed up too tightly one may get down and be unable to regain its feet, with the result that it will soon be suffocated by the heat of those above. The first sign of this, something all sheepmen (especially shearers) should be on the alert for, is sheep jumping up in the pens. As soon as this occurs someone should be over quickly to make an inspection.

In most cases, however, owners are reluctant to pack sheep up because of this very danger. It should also be remembered that sheep can be packed more tightly while the shed is working than when left overnight in a pen. While sheep are moving through the shed and catching pens are constantly being filled, an intelligent 'sheep-o' can do much to improve the shearing of 'sticky' lines without any danger of smothering. For a night pen, on the other hand, a man should be able to walk through the penned sheep with reasonable ease.

CHAPTER FIVE

Crutching

CRUTCHING is an important part of sheep farming which should never be neglected, and it is best carried out at two different times of the year. In early autumn, when the rams are put out, it is usual to tup- or ring-crutch ewes. This involves only two or three blows around the tail, leaving the ewes clean for the ram. It is not a good policy to take too much off at this crutching, as it spoils the main line of crutching wool later in the year. Then in mid-year the main crutching is completed, at any time from two months to a week before lambing.

As conditions differ so much, it is not possible to lay down any hard-and-fast rules about crutching. In some very wet areas sheep are almost half shorn at crutching time; in dry areas very little is taken off. I am of the opinion that the main, mid-year crutching can be overdone as well as underdone. No more wool should be taken off than is necessary to keep the sheep clean, and to allow the lamb to feed and find the teats easily, for heavy crutching can spoil the fleece when it comes to shearing time, causing a large quantity of half-wool pieces. Sheep that are crutched heavily are usually fat-lamb ewes, the sheepowner being not as wool-conscious as the man who farms for wool and breeding.

There are two methods of crutching. One is the old traditional method of coming in from the hock out to the tail, with the sheep turned the same way as in shearing. This is known as the Fantail style. The other is the modern method of coming in from on top of the tail (*Figs. 24* and *25*), and running out to the hock. With this method the sheep are turned the opposite way to shearing. It has been called the Boomerang style.

I believe this last style is the best, having practised it for many years. In it the sheep stays in the one position. The Fantail crutch, involving two positions, is physically harder, and in operation is slower. Also, in my opinion, the modern style of crutching makes a better job, for the wool around and over the tail is cut clean from the back, whereas with the older, Fantail method, coming in towards the natural lay of the locks, there is a tendency to lay or fan the wool back more than when making a clean cut. I agree that when a sheep is done by the old method and stands up, it looks good, with the wool fanned and standing up all round the tail and hindquarters, but it still looks no better than a sheep crutched by the modern method. If you look at the two styles a fortnight later there is no doubt which is the better, for the wool that was standing up in the Fantail style has then dropped down and in many cases, on long-woolled sheep, the wool tips are hanging down over the tail. The modern method still presents the clean cut which gives a practical neat appearance.

For reasons of space only the modern style of crutching is set out here, as being the more practical one to use. It is acknowledged, of course, that the Fantail method is still widely practised and is preferred by some flockmasters.

There are three types of crutching:

1 On high or dry country, where little wool is taken off, the shearer should make one blow above teats, take the crutch out and take a little more wool than will cover the hand from under the tail.

2 On average country varied amounts of wool are taken off in the standard crutch. Here, I believe, there should be three blows above the teats (actually it is only two and a half as the first blow is on the angle to make a clean cut of the belly wool). Take the crutch out, and, starting approximately three inches above the tail, follow an even half-moon pattern to each hock (*Fig. 26*). Sheep that require eye-wigging have the topknot removed, and wool is also taken from below the eyes if required.

3 In very wet areas, where there is high productivity and rapid growth in the spring, the whole, or at least half of, the belly

wool is taken off, both flanks are shorn right out (so that you can look straight through from one flank to another when the sheep is standing up), and the same hindquarters as in type 2. The head of every sheep is shorn almost back to the ears.

This third type of crutching is not practised nearly as much as the first two types, but is seen in quite a few fat lamb areas in New Zealand.

The price that should be paid for crutching is very hard to assess owing to these different requirements, but a fair basis would be one-third the price of shearing for the standard type 1, a little more for type 2, and quite a lot more for type 3, as it is taking the flanks out that requires time and effort. In fact, type 3, taking full belly-wool off, is worth half the price of shearing.

The Standard Crutch

1 CATCHING

The catch is the same as set out for general shearing, differing only as the sheep reaches position on the board, when its legs should be facing the catching door and it should be lying on its off hip (*Fig. 21*). It will be found with practice that with good footwork this starting position can be reached every time. For crutching it is not so important to be in the right starting position as for shearing, as a man is still able to crutch proficiently slightly back or forward of the machine. However, aim at being in the right place, which is straight opposite or a bit forward of the down-tube. It should be noted, however, that when the sheep does not land in quite the right place it is crutched where it is, and not physically dragged into position.

2 THE BELLY BLOWS

It will be seen in *Figure 21* that the sheep is lying away from the machine, not on its back or tail, but on its off side. The sheep is held by lying in against the left leg and knee, while the left hand comes down to part the wool slightly for the first blow. *Figure 21* shows this, as it is important for a quick clear start. The first blow is put in on an angle out from the sheep (with the bottom tooth of the comb being on the sheep, the top tooth up a bit from the sheep) and follows straight across the belly. This angled blow makes a clean cut and break of belly-wool, leaving nothing to drop off out in the paddock.

The next blow is straight across beside the first, and the third blow is on top of the teats and should be put in carefully to avoid cutting them. The sheep is still in the same position.

Shearers will find that on many sheep, instead of this single

FIG. 21. *The start of the full crutch for lambing. 'A' marks topknot wool removed from wool-blind sheep*

FIG. 22. *Inside the crutch (a)*

third blow to clear the teats, it may take two or three short blows to make a clean job and avoid any danger of cutting the teats.

3 INSIDE THE CRUTCH

This is done just as set out for general shearing, preferably with four blows, only with the sheep still leaning away from the machine and not in towards it as in shearing. The left hand pushes in the near flank, forcing the near leg up and out, allowing the handpiece to clean inside it. *Figure 22* shows this clearly. Note also the shearer's body position. Care should again be taken here to avoid cutting teats by placing the fingers of the left hand over the teats in young sheep while the wool is cleaned from this place. *Figure 23* shows the left hand in the opposite flank to straighten the off leg while the handpiece cleans inside it.

FIG. 23. *Inside the crutch (b)*

FIG. 24. *Entering over tail on outside of crutch*

4 THE BACK OF THE CRUTCH

From the last blow in the crutch the handpiece comes outside the sheep and starts in square above the tail (*Fig. 24*). Note also in *Figure 24* that the sheep has been allowed to tip more on its side, and the shearer's body has taken a new position, with the left hand pressing on the near hindquarter (and not the paunch) to keep the leg straight. From here *Figure 25* shows the next blow finished around in a circular even sweep to the near hock. This is also an angle blow, like the first one on the belly, it being imperative to concentrate more on making a clean finished edge to the crutch, than to try to cut a lot of wool off with this first back blow. *Figure 25* also shows the next blow completed. This is put in flat on the sheep around to the near hock, allowing it to run around underneath the hock, finishing to the toe, thereby giving a clean inside leg. Now, as in *Figure 25*, clean across under the tail to inside the other hock, finishing out on the off side of the tail. When cleaning the tail it is imperative to run the comb square across the end of the tail to make a clean finish.

Figure 26 shows the last blow going in, this being an angled circular blow to match the first one on the back. Note in *Figure 26* that the shearer's body is now right down over the

FIG. 25. *The back of crutch (mid-way)*

FIG. 26. *Completing the crutch*

sheep, the left foot having slid back slightly and the left forearm is across the sheep, with the left hand clasping the wool on top of the tail, where exerted pressure and leverage easily turn the back end of the sheep up. *Figure 27* shows the rear view of the sheep when crutched. Note the even-patterned clean finish, which is essential to good crutching.

By looking at the sheep in the pen when finished, a shearer who is new to crutching will soon find from which side he is taking off too much or too little. Rectifying this will give an even crutch.

It has been said that with this style of crutching a shearer is likely to ham-string the sheep. Any shearer can soon prove this wrong, as with average reasonable care he will crutch by this method many thousands of sheep without ham-stringing a single one of them. Ham-stringing on adult sheep could only happen if, when a sheep was kicking violently, the shearer jabbed his handpiece forward on the hock at the same time as

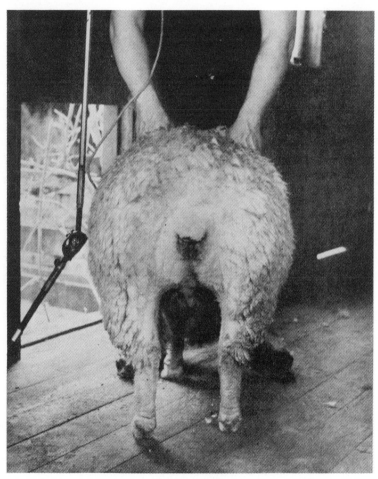

FIG. 27. *The completed crutch*

FIG. 28. *Eye-wigging*

the sheep's leg was moving back. There is a slight danger of ham-stringing on daggy hoggets or lambs, but if reasonable care is taken, it will indeed be unusual to ham-string a sheep with this style of crutching.

Wethers are crutched in the same manner. No wether should

be crutched without being pizzle-ringed. It is again essential that the left hand should part the wool here, and extra care needs to be taken to be sure of not cutting the pizzle. Wethers require only the back of the purse to be done, and do not require so much to be taken off the back as a ewe.

In crutching ewe lambs, or two-tooth ewes that have not been with ram, only one blow inside the crutch is required. On a struggling ewe lamb there is likely to be danger to the teats if you try to clean right up to them. This one blow should be kept a safe distance back from the teats.

The blows set out in this description of crutching are the minimum, and for learners and average shearers an extra blow here or there will be necessary to complete a good crutch. Further, many sheep are 'sticky' at crutching time, and for any shearer this sticky crutching involves a larger number of blows.

Eye-wigging

With the improvements and advances made in breeding sheep in the endeavour to put as much wool on them as possible, many of our sheep have become wool-blind, i.e. wool grows over their eyes so that they cannot see. These sheep are usually eye-wigged at crutching time. A lot of this eye-wigging is grossly overdone, the head being practically shorn. The wool on the forehead, known as the topknot, is the wool that hangs down over the eye, blinding the sheep. The more the wool below the eye grows the longer it gets, and the more it will hang down away from the eye. For this reason I recommend only the topknot being removed on those sheep that, although being partially wool-blind, have a reasonably clear eye. There are, of course, some very wool-blind sheep that do require wool to be taken off below the eye. *Figure 28* shows a sheep being eye-wigged.

If eye-wigging is required with Boomerang style crutching it should be done as the first step, for this makes possible a much smarter switch off and exit of the sheep than if it is done after the crutching is completed. With Fantail style crutching, eye-wigging is done last, after the crutch is completed.

It has been my experience that a wool-blind sheep is not as

good as a clear-eyed sheep, in constitution, in vitality, in breeding, and in rearing of lambs, and is also usually quite a bit below its clear-eyed neighbour in all-round production. Agricultural colleges have carried out tests in this country and wool-blind sheep on exact tests have been found to fall below the production figures of the clear-eyed sheep.

Catching for a Shearer when Crutching

Although a shearer should catch his own sheep when crutching it is common to see someone catching sheep for him. If they are not handed to the shearer properly he gains no time, for he has to pull the sheep into position with one hand, and as this takes considerable effort he often tries to crutch the sheep well out of position. In any case, if someone is catching for him he gets little chance to straighten his back, and it soon tires.

The correct way to hand a sheep to a shearer who is crutching, is to bring the sheep across in front of him, passing the shearer with your back to the porthole wall, and then, when in the right place, to let it fall in towards the shearer. By this method the sheep will drop into the correct position. This is much better than having it on the board alongside the shearer, in a position from which it has then to be hauled bodily across into position.

Crutching is hard work. As there are few blows they do not take long to master, with the result that the harder a man works the more he will do in a day. It is rather surprising to work out the weight of sheep a good man has handled in a day's crutching.

The best way of working when crutching is to go at a good smart pace for half an hour and then have a five minutes' blow. Rhythm, snap, balance and smart footwork enable you to catch and handle the sheep better and much more easily than if you get stale and tired on catching, allowing the sheep time to resist and fight against you. Handling is the biggest part of crutching, and a good job should be done when the sheep is on the board, as little time is gained by slumming these few blows.

The last important point in crutching is sweeping the board. When each sheep is crutched the board around the shearer must be swept clean, for if the new sheep lies on wool already on the board this wool will then stick to its side and back, and the sheep will walk out of the porthole carrying with it several locks of wool. This is why a lot of tally pens and yards holding freshly-crutched sheep present the appearance of a fall of snow, and very often shearers are blamed for not making clean blows when it is the fault of the 'Broomie'.

CHAPTER SIX

Gear: Grinders: Sharpening Cutters and Combs: Handpieces

HAVING THE GEAR right is just as important as shearing in the right way. You never see a good chopper without a good axe or a good shearer without good gear, which is the foundation that all good shearing is built upon.

The Grinder

The first point is the grinder. It is impossible to grind gear properly on a grinder that is not up to the mark.

A double-ended grinder is preferable, so that discs can be changed around each day. As the outside of the disc is travelling much faster than the inside it grinds gear more on one side that the other. This fault is remedied by changing discs around every one or two days.

It is essential that a grinder run at a good fast pace. A common fault in a grinder is running too slow. (Makers of the various machines will advise the correct pace for their model.)

An absence of vibration (on a belt-driven grinder, check bearings) is essential to good grinding.

The best type of installation is a concrete block going down through the floor into the earth, with a three-inch or four-inch wooden block on top to which the grinder is fixed. Any good grinder is worth this installation, which makes for an excellent set-up.

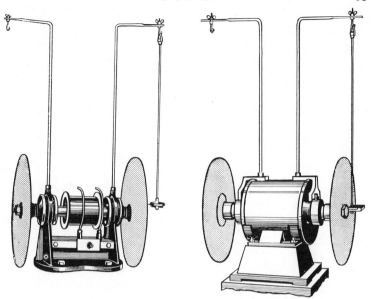

FIG. 29. *Grinders.* LEFT: *Belt—spindle drive.*
RIGHT: *Independent motorized drive*

A magnetic holder is preferable to a spring holder, i.e., it is better to have one in which the cutter and comb are held by pins and magnet instead of pins and spring. In districts of small farms where there is a lot of individual shearing, a shearer is advised to carry his own holder.

If a shed is putting through a big number of sheep, two clamps are necessary: the extra one will soon pay for itself. Sometimes plates are not true, and these should be returned to the maker and replaced.

FIG. 30. *Position of operator when using grinder*

The grinder should be installed in a good accessible place, and not stuck in a corner, or behind a bale of pieces or bellies. Bad placement is dangerous, and a shearer should be able to stand square on to the grinder for good grinding. I once saw a 'ringer' leap out of the grinder room with a mighty yell. He had been grinding in a dark corner amongst a lot of other shafting, and as he turned around the disc ground through his pants into his flesh. His leap amused the onlookers, but it was a very painful injury.

FIXING EMERY PAPERS TO DISCS

The first point is to have papers in good order. Emery papers should be stored in a clean dry place, not rolled up, but lying flat down. (A hot-water cupboard is a good place.) They must be kept away from dampness.

For combs, use a coarse paper. The coarser paper is the better, i.e. a No. 3 is better than a No. $2\frac{1}{2}$. For cutters use a fine paper, but a coarser fine is preferred, i.e. No. $1\frac{1}{2}$ is better than a No. 1.

The discs must be clean and quite free from grease, rust, oil, etc. To keep them clean it may be necessary to wash them in hot water. When the old paper is taken off the disc should be soaked with cold water, and the old glue completely removed. An old cutter does this job well.

Having the disc clean, the next point is to make sure the face of the clamp is clean. If it is a bit dirty, rub the dirt off with old emery paper.

It is essential to have good glue to hold the emery paper on the disc and a good brand should be used. If the plate is hot from washing in boiling water, allow it to cool before applying glue. Apply a thin even coat of glue to the disc, spreading carefully and evenly over the surface, and rubbing well in to the grooves or pores of the disc. Be careful not to use too much glue, as it will come through the paper and form a hard lump. Cleaning the glue brush on the cloth back of the emery paper is a good practice as it prevents blotting the glue on the disc when clamped. Now put the paper on the clamp, then the disc, and screw down tight with the grinder spanner.

It is necessary to leave on the clamp for about twelve hours, but if it is wanted in a hurry, it can be chanced much sooner than this. The edge should be trimmed around the disc with any convenient cutting instrument before taking off the clamp.

If it is a wet morning, or if the shed is full of sheep with humidity high, it will be found that the emery is soft. Do not grind on the emery while it is like this, but endeavour to dry it out with some heat, or if there is any sunshine, with a few minutes in the sun.

THE SETTING OF THE GRINDER

There is an adjustment on every grinder to shift the pendulum of the comb or cutter holder in or out.

Put a slightly used cutter on the holder, hang it from the hook at top, and place the cutter against the disc, checking the distance. When the cutter is in position and against the disc the distance from the pendulum arm to the disc must be the same all the way down. This means that the holder is hanging perfectly plumb with the face of the disc when a good thick cutter is on. If it is too far in at the top it will have a tendency to take the points of the cutter; if it is too far out it will take the heels off the cutter.

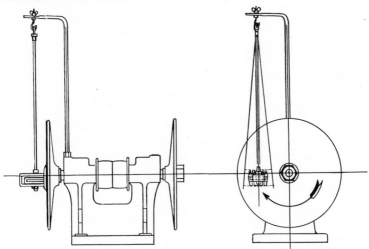

FIG. 31. *The position of holder pendulum on grinder.* LEFT: *Pendulum arm must be parallel to face of disc when cutter is in position.* RIGHT: *Correct position of holder slightly below centre line.*

The next adjustment is the length of the pendulum. The holder can be raised or lowered on all grinders. The correct position is a fraction below the centre of the spindle so that the sparks are running straight out at the point (perpendicular). If it is too high the sparks will run around or across the teeth; if it is too low, they will run across the other way. It is better to have the holder a fraction on the low side, as the comb teeth sit higher on the holder.

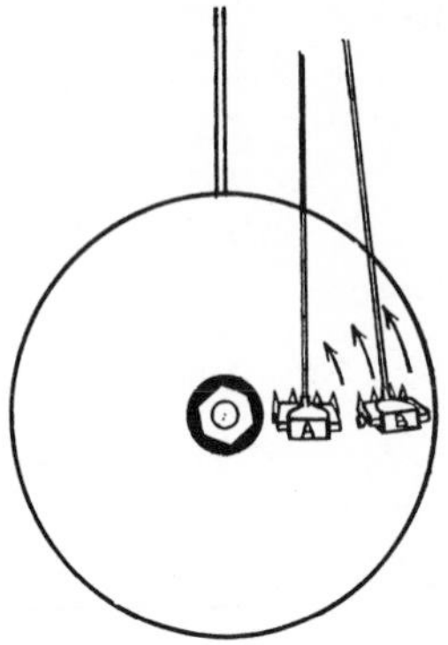

FIG. 32. *The grinding arc. Note that sparks run straight up from tips*

The final adjustment is made with the holder hanging plumb down so that the inside edge of the holder is about half an inch or one inch away from where the flat centre of the plate starts.

Grinding Cutters

When the cutter is put on the holder, make sure it is hard back against the bar up on the top of the cutter. It is important that the heel of the cutter should touch the disc first. Magnetic holders are set so that this happens. It is possible, however, if the pins are tight or the magnet not strong, for the cutter to lean out slightly from the holder bar so that the tips touch the disc first. If this happens the cutter is tipped, and then it will not cut and it will take a lot of grinding to grind the points out again (heels first slightly, and then the points).

It is better to grind a cutter for twenty seconds with a light finger pressure than for ten seconds pressing hard. Strong pressure will heat the cutter, burn it, and not allow the emery paper to cut to its best advantage. Stand squarely on to the disc, and make sure the cutter goes on the emery squarely and comes off squarely, i.e. that all teeth of the cutter go on and come off simultaneously. This is important.

If no sparks can be raised by a reasonably light pressure, the emery is worn out and needs replacing. You will never sharpen gear on a worn emery and probably will do the gear more harm than good. Watch the sparks flowing from the cutter. When first it is put on the disc, very few sparks will be noticed coming from the points; then, as it is ground, sparks will thicken up at the points and flow from each point of the cutter in an even flow (like a shooting star). This means that the cutter is ground out to the tips (the most important part of a cutter) and is properly ground. If there is even a small gap in the flow of sparks from the point the cutter is not ground out to the tips, and is not properly ground. You can tell a lot about the sharpness of a cutter from looking at it, but the sparks on the grinder are a true indication. However, either way it is essential that the cutter be ground right out to the points.

Backs of cutters should be checked and levelled and new cutters backed off, by taking them in the finger tips and putting them back on to the emery. It will be found that fifty per cent are not level across the back: grind these until the backs of all points (i.e. where the fork presses on the cutter) are perfectly level. On some cutters the points are found to be needle sharp: these should be slightly squared off on the grinder as a square blunt point has more metal to stand up to cutting than a needle point. This squaring off can be overdone, and cutter tips should be squared off just lightly.

In both cutters and combs, move from one side of the disc to the other, using the whole of the cutting portion of the emery from the inside edge to the outside edge and no further. Take the cutter or comb off and put it on, on the inside of the disc, (the pace of disc here is not as fast as outside) where the pendulum arm is hanging perpendicularly. If, after the grinder is set up as above, the combs or cutters are still grinding at the heel more than at the tips, increase the distance between the pins (see A, Fig. 33) and the magnet bar of the holder; if, on the other hand, they are grinding more at the tips, then decrease the distance.

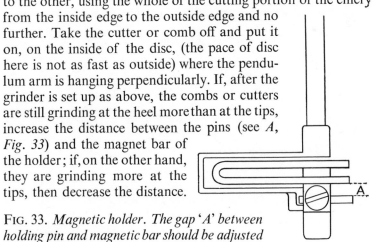

FIG. 33. *Magnetic holder. The gap 'A' between holding pin and magnetic bar should be adjusted as explained in text.*

Grinding Combs

A comb is the most important part of the gear, and the right condition of the comb is the foundation of all good shearing. Many shearers are handicapped by shearing with a comb that is not up to the mark.

The first point is grinding the comb. It seems quite a simple procedure to put a comb on a grinder and to move it backwards and forwards. However, grinding a comb goes further than this.

Exert slightly stronger pressure when grinding a comb than a cutter. The most important point is to know when a comb is sharp. Looking at it face on does not give very much information about its sharpness. As the edges of the teeth do the cutting each tooth must be inspected as to its sharpness. The way to do this is to hold the comb on its edge, with the light (from a window or other light) reflecting on the edge of the comb teeth (*Fig. 34*). On a blunt comb will be seen a small white hair-line where the light is reflecting on a round edge down the edge of the comb tooth. When the comb is sharp these white lines will not be seen, and the comb should be ground until they are all taken out.

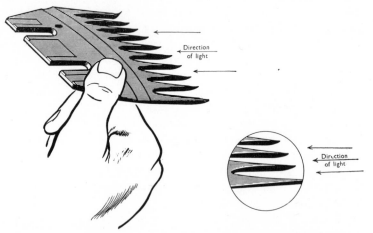

FIG. 34. *Inspecting a comb for sharpness. A sharp comb, held so that the light falls along the cutting edge of the teeth, does not show a white line. The inset shows how a blunt edge will reflect a strip of light.*

A comb properly ground will shear with a light tension and cut perfectly. When, after doing ten or fifteen sheep, the cutting starts to dull, change a cutter rather than screw the tension down and carry on. Screwing down will only put white lines on the comb. Once, when changing my gear, a fellow shearer said: 'Changing your cutter, Godfrey?' I said, 'Yes. Why?' He said, 'Boy! If I could get my gear to cut like that I would never change it.' That was why he did not get his gear to cut well. You do not run and run gear until it will not cut any more, and even before 'smoko' it is better to change a cutter for one or two sheep than to screw the tension down, a procedure which dulls the comb very quickly.

A comb needs plenty of grinding, and it must be sharp. It takes a sharp comb and a sharp cutter to cut wool off—not a sharp cutter and a blunt comb—a combination which is too often used.

Once I went to a shearer who was grinding gear. He just touched the comb on the grinder for a moment and then took it off. I said, 'Why don't you grind your gear?' He replied, 'You wear it out too quickly.' Shearers should not be afraid to grind their gear. It is better to wear it down and have good sharp gear, than to keep it thick, but blunt and useless. Sharp gear cuts a lot of wool off and earns the shearer a good remuneration for the time it is used.

Doing Up the Points of a New Comb

This is important, and it is rather hard to set down a definite pattern, as the points of combs differ, just as sheep differ. In fine-wool sheep one shears with a much brighter point on the

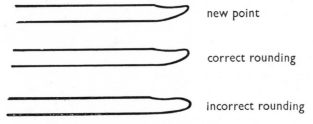

new point

correct rounding

incorrect rounding

FIG. 35. *Points of comb teeth. The centre diagram indicates the correct shape of point*

comb than on crossbred sheep. However, the basic principle
is the same, but remembering also the rule—brighter on finewools and blunter on crossbred sheep. The point of the comb
should be like the end of one's finger or the end of a common
table knife—i.e. perfectly rounded as in the diagram and not
blunted off with oblique angles (*Fig. 35*). This means that the
point is kept high and rounded instead of squared off. Remember that a comb goes on the skin at a forty-five degree angle,
and this type of tooth keeps the point entering the wool with
the round underneath following on the sheep, resulting in a
blunt comb which will enter the wool and not mark the sheep.

The way to do this is take a bit of fine emery and fold it
to make it stiffer. On the first tooth, starting at the top or
point, run the emery around to the bottom or back, blunting
the tooth off all the way round. This will take two or three
minutes for each tooth. When the first tooth is done do the
next, getting every one the same. When this is done, taking the
emery, run on the inside of each tooth up on the tip, not touch-
ing the cutting edge. This will take off the burrs you have put
on. This does not take long, applying only a few rubs on the
side of each tip.

The next phase of experting a new comb is to scrape or rub
the tips on wood. Using a piece of soft board take the base
of the comb in the fingers and rub the comb points backwards
and forwards on the wood (*Fig. 36*). This should be kept up
for at least half an hour of continual scraping, when it will

FIG. 36. *Polishing points. Doing up the points of a
new comb by scraping on a soft board*

be found that the comb points have been polished so that each is smooth, even, and rounded. With points such as these wool has little chance of catching to impair the entering of the comb. This scraping on wood is one of the most important operations in doing up a comb. One sometimes sees shearers scraping combs on prominent places in woolsheds. This is not a good practice, as the wood in most cases is too hard, and it impairs the appearance of the shed. The easiest and best method is to sit down in a relaxed position and scrape the comb on a piece of soft board between the legs. The reason for using soft wood is that with points sinking into the wood, the sides of the comb teeth will be polished as well as the tips. A final polishing on the wall of a used motorcar tyre completes the doing up of a new comb.

One cannot expect a new comb to shear tough sheep, shear lambs, and do crutching: it must be remembered that it is new and thick.

An experienced shearer always breaks in his new combs on the better shearing, and when they are half worn down or ground out to the tips, they are put away for the early or sticky shearing. If one wants a fine comb, it is not necessary to fine it down with grinders and trimmers—give it plenty of use on average shearing and as it gets thinner the teeth will get finer. It will be found that when a comb done up as above, does wear down, the points, although thin, will still have the desired rounded edge. When the comb is ground out to the tips (i.e. when the scallop has disappeared and each time it is ground the comb touches the emery right out to the tips) it will be found that after each grind the points sharpen slightly and need to be lightly rounded with emery paper before being used for shearing. This takes only a few seconds.

The secret of a good comb is that, being properly ground, it has sharp teeth edges showing no white lines, and has blunt rounded points that will enter the wool. One often sees square, badly done-up points that will not enter the wool, thereby greatly affecting the shearing. Opposite to these are points that have had little attention, which may enter the wool easily, but are really dangerous and are the cause of cutting sheep. No shearer can be proud of this type of comb or of its resultant job. Unfortunately, one sees far too many of these sharp-pointed butchering combs. I trust that the experting as set out in this text will be a practical guide and help to all, as good gear is the foundation of all good shearing.

Fixing Cutter and Comb to Handpiece

Even when cutter and comb have been put into proper condition they still have to be fixed correctly to the handpiece. First, screw back the tension nut (on the handpiece), i.e. lighten the tension. Now, with handpiece upside down, put the cutter on the fork and slide the comb on. With the comb teeth resting against an upright, lightly tighten the screws on the handpiece. When final adjustments are made, see that they are good and tight. The essential point now is the amount of lead to give a comb and cutter, i.e. where the point of the cutter is on the comb. On lamb shearing, crutching, early or sticky shearing, no lead is put on, as it is essential that this type of wool be cut as soon as the comb enters the wool. This means that the point of the cutter is put right up to the start of the cutting edge on the comb. It should not go right to the tip, for there is a scallop on comb tips for the first quarter of an inch, and this part is never reckoned as a cutting edge, even when the comb is ground out to the tips and the scallop disappears.

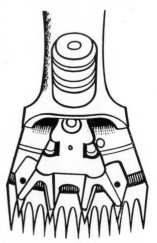

On good open shearing, when the grease is up, and the comb enters easily, give the comb a lead on the cutter of one-eighth of an inch, i.e. the points of the cutter should be one-eighth of an inch back on the comb from the start of the cutting

FIG. 37. *The position of the points of the cutter in relation to the front of the cutting edge of the comb (at the base of the scallop) indicates how much 'lead' has been given the comb.*

edge. This is done by sliding the comb in or out on the screws. It will be found on this type of sheep that the skin is much thinner and more tender, making it easier to cut and mark sheep. Giving this lead allows the teeth of the comb to ride the skin wrinkles before the cutter reaches it. If the cutters are well worn and will not come far enough out on the comb when the comb is hard up, back the comb off on the grinder before renewing the emery paper, allowing it to slide back further.

When adjusting gear it will be found that the comb can be moved slightly sideways. The comb should be adjusted so that the cutter crosses the comb evenly and does not protrude at either side.

It is important when setting the lead on the gear, always to adjust it with the cutter first on one side of the comb, and then on the other. If the lead is adjusted with the cutter in the middle of the comb it will be found, when the cutter comes across to the side, that the outside tooth may be over the scallop.

TENSION ON A HANDPIECE

The next point is to adjust the tension. There are two grave errors in tension adjustment. One is to start the handpiece with too light a tension, so that when the first blow goes in the gear bogs and chews, and small balls of wool collect under the

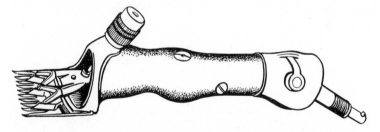

FIG. 38. *A complete handpiece* (*narrow gear*)

cutter. Once that happens a cutter will never cut properly again until it is reground. The other error is to shear with too tight a tension, thereby heating the handpiece and dulling the gear. The way to avoid these errors is as follows:

Screw the tension down until, with the thumb in the back cogs, the fork can just be turned over the eccentric with hard firm pressure of the thumb. Now put on one more notch of tension—i.e. a fraction of a turn on a 'chickenfoot' machine, or the notch on the tension screw of other handpieces. It will be found that you now have the correct cutting tension. If it is too much, then let the tension off; if it is too little, put a bit on. Remember always to shear with as light a tension as cuts well. (However, you must be able to drive without gear-chewing.) Many times I have picked up learners' and even some shearers' handpieces and the first thing I have done, sensing too much tension, is to take tension off. If gear is sharp it does not need tight tension to shear, and with the right tension, the handpiece should not run hot unless, of course, there are special circumstances, such as the machine running too fast, or the shearing of sandy sheep, etc.

Changing Gear

Don't be afraid to change a cutter if it is getting dull. It is better to use one more cutter on a run and enjoy good cutting than to try to make cutters last out. One comb should do a run if properly ground on average clean shearing. A lot of time can be wasted changing cutters, and shearers should train themselves to change quickly. Have the cutter beside you on a nail on the wall, and also see that the handpiece tension screw runs off and on easily (hot water every two or three days will keep this right). On handpieces that use driving pins on forks these pins should be kept short. New ones are too long and should be ground down.

The routine in changing cutters should be: (1) run the tension out; (2) flick the used cutter off into the water tin; (3) concentrate on fixing the new ground cutter on one side of the fork; (4) then slide it around to the other side; (5) click it in; (6) run your tension down. A practical hand takes only a few seconds to do this.

Have a good oil-can handy. Run oil across tips of the cutter, on the comb, and on the centre post and fork eccentric of the handpiece. These are the only three places to oil in a run.

There is no need to pump oil in the back cogs or the short driving spindle, until the break at the end of a run. Even then the back cogs should only be very lightly oiled, as oil will mix with grease and dirt and work down into the back joint, gum the handpiece up and make it run stiff in the back joint. It has always been known that bad aching backs are the cause of a lot of oil being used—as when the handpiece is oiled every two or three sheep. If one's back is aching there is really only one cure—to keep going till it stops aching. To rest it all the time allows it to give more trouble.

It is quite good practice to use cutters alternately, as they will then wear at the same rate, matching each other in thickness. This can be done by sliding them on a wire and taking them off the back.

The notes given here do not cover procedures required for dealing with sandy or gritty sheep or difficult shearing, which are special shearing problems, and are dealt with separately in this text (see pages 47-52).

Handpieces

As there are several different types of handpieces there is not space to set out how each of them should be taken down and experted. However, the makers provide this data. The fundamental rules that apply to all handpieces are set out here.

1 It is not the make of the handpiece, but rather the order it is in that gives the result. All are good handpieces, and a shearer usually sticks to the model he learns with, for he gets used to its balance. Make sure, however, that the handpiece is in good order, as it is the most important part of all shearing plant and must be right. If any shearer is working with a badly-running or worn-out handpiece he is handicapping himself physically and financially, merely to save the small expense of getting parts renewed or done up by a reputable firm. I myself have always walked on to the board with two handpieces going in perfect order.

There have been—and I believe still are—laws that say a shearer cannot provide his own handpiece and gear. This, in my opinion, is a grave error, as the handpiece is just as

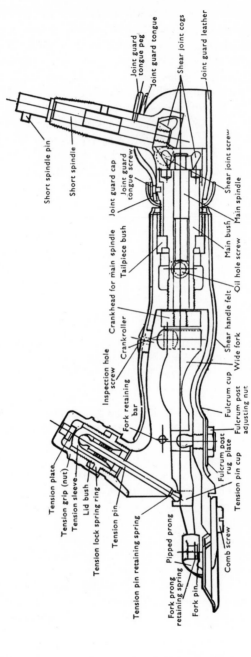

FIG. 39. *Diagrammatic section of a complete handpiece, with names of parts*

much a part of a shearer as the axe of the bushman, the knife of the butcher, and the gun or rifle of the sportsman. The handpiece should be treasured and prized by the shearer just as their special tools or weapons are valued by these others.

2 The oil in the barrel of a handpiece should be checked every run. When filling, do not fill to overflowing, for when the oil warms up it expands and forces itself out of the screw hole, making the handpiece oily and hard to hold. A strong leather or cloth washer should be used under this screw to prevent leakage. Always use good clean oil—not thick grades but still not too thin. A lot of shearing oils are too thin, especially for exterior oiling, and they soon disappear when put on. Grade 20 is the best for all-round work.

3 Make the best of spare time, and at least once every two weeks take the main working parts of the handpiece down and clean and oil them. Handpieces which are closed in around the fork need this more often than the others. Be careful, if taking more than one handpiece down, not to mix the parts.

4 Every two or three months check the driving pin on the small shaft at the back point, and when it is badly worn punch it out and renew. It is better to do this than to wait till it shears off in the middle of a run. This is a tapered pin, and must be punched out from the riveted side. New pins are usually too long, and when put in should be shortened on the grinder disc, for they will cut the short tube if left long. See that they are riveted tight when fitted—the large end is hardened for driving and the other is soft for riveting. This riveted end wants smoothing off with a few light touches on the grinder, for if the shaft is left rough it will seize up in the bayonet joint of the short gut, making it difficult to remove the handpiece from the down-tube.

5 See that the ferrule on the handpiece is kept clean. If it gums up it will give a lot of trouble by not working freely on the short down-tube. Note that a long ferrule is preferable to a short one as it lessens the vibration on the wrist.

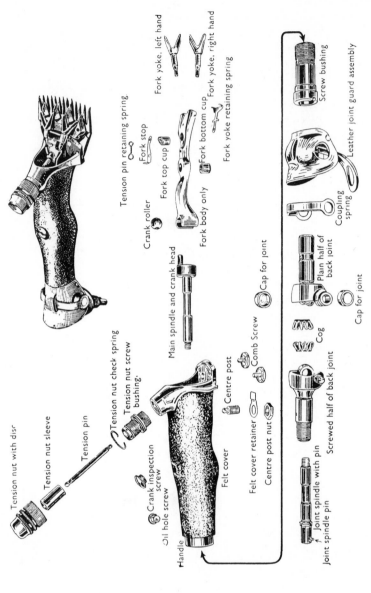

FIG. 40. *The individual parts of a handpiece*

6 A lot of new handpieces come out with felt covers. This is perhaps all right for learners, but accomplished shearers, who have good gear and who know how to run a handpiece, should use a handpiece without a cover as it gives a better and smaller grip, is lighter, and will be found easier to control, resulting in a better touch.

7 If the handpiece is running a bit hot do not run it in water to cool it down, as water has a quick wearing and dulling effect on parts and gear.

8 At the end of every season have the handpiece overhauled. Then worn parts, such as fulcrum cups, cutter pins, fork rockers, driving pins, can be replaced with new ones. A handpiece so overhauled is then ready to start the new season.

9 Never worry about the barrel getting rusty or pitted when stored in the winter, as such a barrel gives a better grip than a shiny new one. However, it is important, that all the working parts and inside of handpiece should be clean and well-oiled when stored. Hot water is the best method of cleaning out grit, dirt, etc.

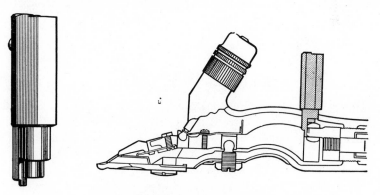

FIG. 41. *Use of gauge for adjusting centre post on chicken-foot type machines*

10 As previously mentioned, watch long fork pins. If thin cutters are used, see that the pins are not bearing on the comb.
These fork pins also come loose, affecting the cutting of the
handpiece, and if left will soon wear the hole in the fork
larger. They are tightened by riveting from the top, the
bottom of the pin being placed on iron with a solid foundation
and the pin then being tapped with a hammer.

11 Most of the 'chicken-foot' machine firms supply a gauge
for adjusting the centre post. This should be used by the
operator when required, after studying the maker's instruc-
tions so that a full knowledge may be gained of how adjustment is to be made. A new handpiece is correctly set before
it leaves the factory and does not require any adjusting.

Storing Gear

When gear is put away for the winter it will go rusty and pit
if steps are not taken to prevent this. There are three good
methods of preventing this rusting up, as follows:

1 The gear can be immersed in dry flour in a tin.

2 The gear can be immersed in oil in a tin.

3 A good smear of petroleum jelly can be put over each article
of gear.

NOTE: Especially watch the combs, as they must not rust or
pit. A rusty cutter can be ground and used quite well, but
not a rusty comb.

Different Types of Gear

There are many different types of gear, ranging from a round
comb with the centre teeth higher than the outside teeth, to a
straight comb and on to a hollow comb in which the outside
teeth are the longest. There are also narrow gear, as used in
Australia, and wide gear as used in New Zealand.

The most important point is always to use the same type of
cutter as comb. Not necessarily the same make, but for
example, a straight comb on a straight cutter, etc. I have seen

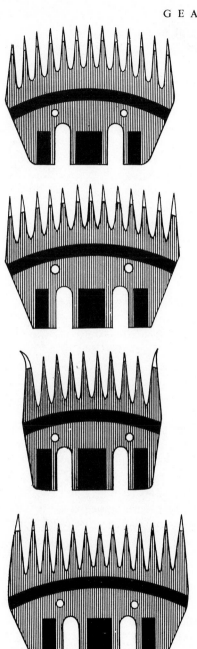

(A) *Standard round wide comb
with its cutter*

(B) *Standard straight wide
comb and cutter*

(C) *Standard straight narrow
comb and cutter*

(D) *Standard wide hollow
comb*

FIG. 42.

and used all types of combs, and prefer the standard straight comb for crutching and early and tough shearing, and recommended its use for learners on all types of sheep. It must be pointed out that the straight comb surpasses all other combs on the inside of a sheep. The improved hollow comb is the best for experienced shearers on all good open shearing, especially second-shear and lambs. Do not round the points of a hollow comb as much as those of a straight comb, and use much less lead on the inside teeth. Do not overbend or turn the inside tooth out on a comb, as this tooth is in the wool all the time working on the blind: a wing really puts a small brake on the shearer's hand. However, it is a good practice to have the *outside* wing bent out, as it is not so much in the wool and does collect the odd 'whisker' that would otherwise be missed. If the comb is purchased with outside tooth not turned out, heat it from the back of the tooth about one-third of the way down from the tip, and when hot enough bend out with a pair of pliers. Don't overbend it into an ugly wing—it needs only a slight even bend (*Fig. 42c*). Make sure when bending the tooth out to turn it out straight, not tipping the tooth forward or back, but making sure it grinds evenly with the comb. When bending, a welding torch will provide a good method of heating if available. (However the biggest proportion of combs should be winged when made—outside tooth only, not the inner one.)

Finally there is the argument about which is best—wide gear or narrow. There is really not the difference that many imagine there is, for the present narrow comb is wider than the early narrow comb. On wrinkly Merino sheep a narrow comb is probably better than a wide one, shears the sheep just as fast, makes a better job and is also easier to use and to manipulate around the wrinkles. However, it is only on these heavily wrinkled sheep that the narrow comb has its value, the wide comb being better on all smooth-skinned sheep, whether fine-wool or crossbred.

It has been said that a wide comb makes more second cuts. This statement is fundamentally wrong, as a wide full blow does not of itself cause second cuts: these, in fact, arise from sheep being out of position, or from shearing with a hurried or uncontrolled hand that allows the handpiece to come off the sheep out into the wool. It is not the gear but the man behind it that makes for second cuts. Use of a wide comb, taking fewer blows to shear a sheep, should result in no more second cuts than use of a narrow comb.

Wide gear is more difficult to control, as with a wider bearing surface there is more tension on the hand and wrist. Furthermore, wide gear does not cut as well, nor run as smoothly, as narrow gear.

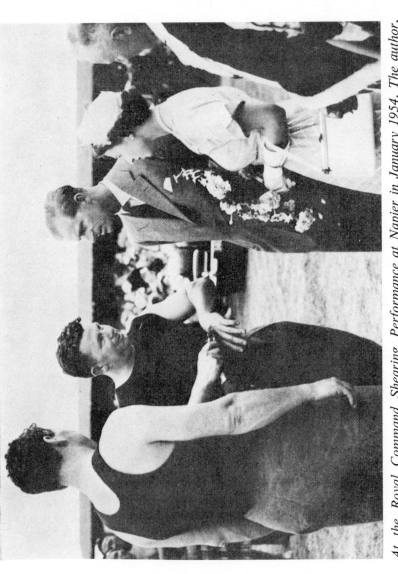

At the Royal Command Shearing Performance at Napier in January 1954. The author, in answering the Duke of Edinburgh's question, 'How is it you don't cut the sheep?', explains the points of the comb by running it up his arm.

The set-up in front of the dais at the 1954 Royal Command Shearing Performance. This photograph shows the author finishing his first sheep. After demonstrating the various points of the shearing, he was partnered by his brother Ivan (standing at left of photograph) in a speed exhibition in which the brothers sheared blow for blow.

CHAPTER SEVEN

Work in the Shed

The Catching Pens

IT IS A COMMON PRACTICE in many large sheds for two shearers to share one catching pen. This can often lead to dissent between shearers, especially if their tallies are close. There are two principles which can be followed and one or other agreed upon by both shearers.

1 For each shearer always to catch the sheep nearest him when entering the catching pen.

2 For each shearer to go for the best sheep available to him, which results in both shearers getting an even cut. This is probably the best method.

When a pen is shared, the pen should be filled before the last sheep is caught, so that the cobbler is not always put on to one man. Also, if there is only one sheep in the pen and both men are finishing together, one will have to wait until the pen is refilled. It is the 'sheepo's' job to prevent such a state of affairs.

A card over each catching pen showing the shearer's run tally is greatly appreciated by shearers. Not only is shearing hard work, but it is also a physical challenge both against the sheep and one's mates. Many a day's shearing becomes a good-natured contest. In this respect a shearer thinks more of the credit he gets for having shorn a sheep than the money he earns by shearing it.

The 'Fleece-o'

The person who works in closest association with the shearer is the 'fleece-o', whose work is to pick the fleece off the shearing board when shorn, to throw it on the wool-table, and also to keep the board swept clean. This last point is very important. The main part of the board to keep swept is between the shearer and the wall, or on the inside of the board. If this part is not kept clean a lot of wool goes out the porthole, especially in chute sheds. Further, it is very inconvenient for a shearer when he goes to pick up his handpiece amongst pieces of wool, and clasps a fistful of wool as well as the handpiece.

To a new hand picking up a fleece is a worrying business, and results in some hopeless antics and tangles. Yet it is very simple, confidence playing a big part. When the shearer finishes a sheep the backbone line on a fleece is usually discernible. Standing over the fleece each side of this line, take the hind-leg part of a fleece in each hand (any shearer will point this out). With some shearers the first hind leg is tucked in under the fleece, and needs turning out. The golden rule in throwing a fleece is that once these hind-leg parts are gripped in the hands, they should never be let go until the fleece is on the table. After gripping the fleece in this manner two common good methods can be adopted:

1 Raise the hand lifting the back end of fleece approximately eighteen inches off the floor, and draw the hands back towards the body, thereby putting a fold or tuck in the fleece. Keeping a firm grip on the fleece where it was first taken hold of by the hind legs, put the hands back to the floor and gather the fleece up in the wrists and forearms. Holding the fleece in this manner, stand in front of the wool-table, then underhand throw it out with a good high even throw, keeping a hold on the hind legs with arms apart until the fleece rests on the table face side up.

2 The other method is much the same, but instead of lifting back the end of the fleece to fold it, the fleece is kept on the floor and the hands draw the fleece back towards the feet in an outward circular motion, gathering the fleece off the floor and throwing in the same manner as in the first method.

However, of the two, I prefer the first method, as the fleece when folded makes a neater bundle in the arms. A neat bundle is essential to a good throw. Do not be afraid to give the fleece a good high throw, treating it as if throwing out a blanket.

A good 'fleece-o' has quick, active, neat movements, is never in the way, but always on the job, and by efficiency can do a lot to keep the board of shearers happy.

Skirting and Rolling the Fleece

The next procedure is for the fleece to be skirted and rolled. Conditions differ so much in types of wool, and in the manner of getting it up that it is not possible to give details about this phase of shed work. The following are the basic points, however, as applying to the biggest proportion of wool get-up:

1 A common fault is for a wool-table to be too high, which makes harder work for the table hands and the 'fleece-o'. The table should be at waist height so that the hands may work at their natural level.

2 In large sheds, the table should be out in the floor where table hands can get all round it. In small sheds, where only one man works on the wool-table, a practical table can be designed against a wall to save too much unnecessary walking and moving around.

3 The table should have good fine planed battens with a good space—approximately one inch—between them to allow second cuts and locks to drop through.

In skirting a fleece use two hands—holding the edge of the fleece in one hand, take off the skirt with the other. If skirtings are pulled off without one hand holding good wool, a lot of this good wool tears out of the fleece and goes in with the skirtings. A proficient table hand soon runs along the side of the fleece in this manner, taking off the required skirtings, no more and no less. Skirting must be done with common sense, but a good all-purpose rule is to take off as little as possible, leaving a clean uniform fleece.

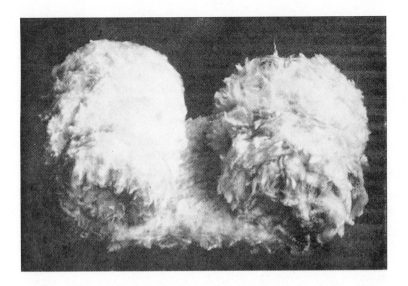

Fig. 43. *Rolling the fleece*

Above: *The store-roll, in which the fleece is rolled from both ends.*
Below: *The shoulder-roll, preferred by the author (see page 110).*

4 After the fleece has been given the desired skirting, taking off stained pieces, seedy dirty wool, brands, etc., it is ready for rolling.

The best form of rolling is that which can be done speedily and efficiently on the wool-table, which will hold the fleece together, and which will present it to its best advantage. The much-used store-roll, where the fleece is folded in from both sides, and then rolled from the ends, has the following disadvantages:

With the store-roll the bulk of the wool showing is back wool, which is not a fair sample of the average fleece. Back wool is often the weakest part of the fleece, and when rolling from both ends it is easy even to exaggerate a minor weakness and mar the general appearance, this being particularly so with table hands who roll too tightly. A presser can also break a fleece rolled in this manner by putting a foot on each side of the roll when forcing it into the press.

It is not suggested that any particular roll will greatly influence a wool-buyer, but the following suggested shoulder roll usually presents a fleece to better advantage, and should give a fairer example of what the fleece contains. Even with the same type of wool the overall presentation and get-up of a wool-clip can give a much better impression when well and properly done, than when the wool is just bundled together.

THE SHOULDER ROLL

When the fleece is skirted, turn one side in by one-third (i.e. one-third of the overall width of the fleece). Now fold again to meet the opposite edge of the fleece. The folded fleece is now rolled from one end, thereby giving shoulder and not back wool as the bulk of the wool presented, and showing and giving a better general appearance. This shoulder roll is simple and fast, it goes into the press in one roll and not as two rolls folded together as in the other method, with very much less likelihood of rolls coming apart or of fleeces being broken.

'Bellies'

Bellies often receive less attention than they warrant, and are usually just baled up and forwarded as bellies. There is, however, a big difference in ewe and wether bellies, and they should not be mixed, as a bale of bulk ewe bellies can be depreciated a grade because of the inclusion of a few wethers' bellies from which urine stains have not been removed. For this reason ewes and wethers should be separated by drafting. It is also a help to shearers to shear one sex at a time. While this is always done in big flocks, in many small flocks drafting is not bothered with, and can result in mixed bellies. If possible, urine stains should always be removed in wether bellies, and even though there is not the time in many cases to do this when shearing is in progress, these bellies can often be put to one side and sorted when opportunity arises. An excellent idea I have seen adopted in some sheds, is to have a platform built above the catching pens where the 'fleece-o' tosses the bellies. They are then quickly put out of the way to await sorting.

Pressing

Pressing wool is a simple straightforward job, but good pressers work without waste movement and without getting in each other's way. This makes them pretty to watch, and also results in a large quantity of wool being pressed in a day. The following are a few points to observe on this job:

1 In loading small fleeces tramp them in layers of five, one fleece in each corner, one in the middle. With larger fleeces place one in each corner only.

2 In tramping a press, concentrate on the sides and not the middle, as the sides will bind down and hold, only the middle springing up and down.

3 Always press the bottom half of the press the tightest.

4 Try to keep the bales an even weight by counting fleeces, remembering that light bales result in higher costs for bales and freight, while wool packed too tight does not open up so well in the stores. A good medium should be aimed at.

5 In those tip-over presses that have iron pins to hold the wool in, do not forget to remove the pins before pumping down. I have seen many heated moments around presses through pins being left in.

6 When the bale is pressed, see that it is numbered and branded neatly. This takes no longer than doing the job in a careless manner.

7 When the bale is branded enter up on the bale sheet its number, grade and contents. If these two last points are not watched, a big muddle can be made of a wool clip.

8 In countries where bale tops are sewn, as in New Zealand, a simple lock-stitch or running-stitch should be used. One length of twine doubled should do one side. Much time can be wasted in sewing, but with a good needle and a fast lock-stitch it should not take long to stitch a bale. It will be found much easier to work always in one direction, either to the right or to the left, according as the sewer is left-handed or right-handed.

The 'Sheepo'

The 'sheepo' is the man who fills the catching pens. He gets this title from the fact that when a shearer catches the last sheep in his pen, he gives the call of 'sheepo'. Shearers usually like to keep in with the 'sheepo' in the hope he will give them a large share of the good sheep. However if a 'sheepo' wants to stay popular with the board of shearers he will be fair at all times—he will not last long otherwise.

The 'sheepo' should never be caught putting sheep in the catching pen when a shearer comes in from the board. A shearer cannot be expected to wait. If a 'sheepo' is stuck for time—that is, if he has not had a chance to fill the pen before the shearer has nearly finished his last sheep—he should whip one or two sheep into the pen and put the gate down until the shearer has caught a fresh sheep. He will then have ample time to fill the pen. He should not fill the pen too tight, with all the front sheep facing the door, leaving the shearer no chance to turn the sheep round to catch it. I have worked with some good 'sheepos' who always had one or two sheep stern first to the catch door when they filled the pen. The 'sheepo' should make as little noise as possible, for banging and shouting get on a shearer's nerves, and they do not cause the sheep to load any better.

When he comes to the end of the mob, with all the sheep of the mob in the catching pens, he should sound the call 'all on the board!' to let the shearers know that the finish of the mob is coming. They can then work out catches, etc., and also avoid boxing mobs. A 'sheepo' should keep in touch with the 'Ringer' (or the 'Rep') as to how the shearers want sheep packed up, etc. This probably sounds a bit regimental, but it pays to remember that the shearer has the hardest job in the shed and anything that can make things easier for him is worth-while.

The Tally Clerk

When 'runs' are finished the tally clerk should be ready to tally sheep out, for when the run is finished, a shearer likes to know how he has been going. Sheep should be carefully and correctly tallied, as shearers know when they are one ahead or behind their mates' tally. I have seen many heated arguments about this, and on checking it is sometimes found that a sheep has jumped out. Correct tallying—and giving a shearer the benefit of the doubt—will save at times a great deal of friction.

A very good practice is to have the tally board hanging up on the wall where all the shearers on that particular board can see it, allowing them at any time of the day to know how their tally is progressing. It can be very exasperating to a shearer to have to ask the tally clerk how he is going, and have him produce the book from some hidden corner, or from a back pocket.

Tally pens should be large enough, where possible, to hold one run of sheep.

CHAPTER EIGHT

Care and Repair of Shearing Machines

THREE COMMON TYPES of shearing machinery are used—shafting plants, single electric units, and portable plants. The largest number of machines at present in use would be shafting plants, but single electric units have become very popular in recent years and form a large proportion of new machines installed. While there are few portable plants in comparison with the first two types, they are found to be essential for shearing in some places, and are also used on some big runs for crutching and for dagging.

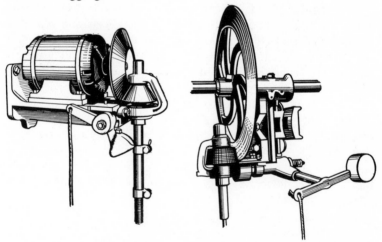

FIG. 44. *The two main types of drive for shearing plant.* LEFT: *The single electrical unit.* RIGHT: *A shaft-driven unit.*

The following are a few fundamental rules relating to the care of the several kinds of shearing plant:

SHAFTING PLANTS

The various makes of shafting plants have much in common, although they differ in some points of design. In all of them proper lubrication is essential. The oil used must be of good quality, clean, and of the grade specified by the makers. Where grease is required, it must be clean grease. Waste oil should not be used, and leaving tins of grease open, allowing grit and dust to get in, will lead to trouble with ball-bearings. Three different types of oiling systems are used with shafting plants—the wick feed (old type), the bottle feed, and the ring oiler system. Although the servicing of each differs slightly, the owner of any particular plant will understand the servicing of plants of his own machinery. The important point with all types is to see that oil reservoirs are kept filled and clean, and that plants are regularly serviced when shearing is in progress.

ELECTRIC PLANTS

Electric plants are either direct-driven or use the more popular friction drive with a friction wheel and cone. Very little lubrication is required on an electric plant, apart, of course, from the down-tube, which requires constant oiling. With most of the modern plants sufficient internal grease is put in at the original assembly to last for approximately ten years. After this time the grease should be renewed, preferably by the makers of the machine.

An electric motor must not be over-lubricated or damage will be caused to the carbon brushes. When an electric unit gives internal motor trouble it should be returned to the makers for repair, as the motor is of a special type. Grease cups are provided on most of these units, and should be given a turn every one or two days when the unit is being used.

LUBRICATING DOWN-TUBES

From the friction wheel or driving cone down, all plants, whether electric or shaft-driven, are much the same, and it is not possible to give the down-tube too much lubrication at the end of every run. Oil should be put in the several oil holes

that are provided for lubrication. Without this essential oiling
the down-tube drive will soon run hot, especially at the elbow
joint, and the drive will start to chatter and vibrate. It does
not do a down-drive any good to let it get hot in this way.
Also, it can delay a shearer's work.

Occasionally wool will get into the elbow joint or short down-
tube. When this happens the gut should be taken down, all
the strands of wool should be cleaned out, and the gut well
oiled before replacing. One strand left in will cause heat.

At times a gut will break. Every shed should have spare guts
on hand so that such a mishap may not cause a long hold-up.
New guts can be stored for protection in a piece of pipe, for
rats have a special taste for new gut cores. Guts can be repaired
by using a splicing ferrule. When repairing a gut see that the
broken ends are paired from the frayed end right back to solid
gut. When they are right up in the ferrule, a small nail of the
right diameter must be used as a rivet. A steady hand is needed
to drive it through the holes at the top and bottom of the ferrule,
but this job is usually soon accomplished. See that these nail
ends are flush and not protruding from the ferrule, or they will
cause friction on the tube. It is bad practice to repair a single
broken gut by splicing it back on itself, as it will then end up
too short. You need two broken guts to make the length
required for a repaired one.

CORRECT LENGTH OF GUT

The correct length of a gut is shown when, with the long
and the short gut connected, they hang so that the short gut
swings just clear of the floor. This length is important, and
when machines are installed, or repaired guts put in, this correct
height of the dropper should be strictly observed—i.e. one-
quarter inch above the floor. If the height is either more or
less, the shearer will be seriously handicapped. If the gut is
too long the down-tube will pivot on the floor when the handpiece is connected, and when the machine is switched off the
handpiece will slide all round the board.

The worst fault, however, is to have the gut too short, for
then, on an average big sheep, a shearer will not be able to
reach the long blow and the last side comfortably. I stress this

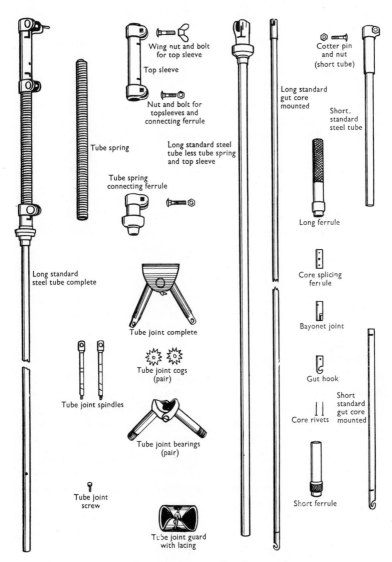

FIG. 45. *The parts of a down-tube*

point strongly, for I have seen hundreds of plants with droppers anything from three inches above the floor to two or three inches below (i.e. standing out two or three inches from its right position). Remember that the height is not measured by the down-tube, but by the gut or drive. In any case the end of the short down-tube should be approximately flush with the bayonet joint of the short gut.

New gut cores are stiff and dry, and when assembled before the down-tubes are put on they should be thoroughly greased, the grease being rubbed well in. However, do not put excessive grease up at the top, where the core is attached to the friction spindle catch, for grease around this catch has a tendency to slide the locking pin up when the long down-tube is pushed on.

This is why at times the down-drive comes unattached and collapses. It is essential when assembling the long down-tube to give it approximately three-quarters of an inch of end play— i.e. when the whole down-tube, comprising both the long and the short parts is assembled complete, the top wing nut should be slackened, allowing the long down-tube to be lowered as far as it will go. The long tube is then raised the required three-quarters of an inch and the wing nut up at the top is tightened.

DRIVING CONES

Leather friction driving cones develop small flats on the surface and cause very noisy running. This is usually the reason for a plant to be 'kicking up a din'. Such flats are easily removed with a special tool known as the cone trimmer, which almost any shearing machine firm will supply. All driving cones are adjustable, and should not be given a big clearance when in the off position or they will slip through lack of pressure. In close humid weather, especially in a shed that has been full of sheep overnight, cones and friction wheels will get damp and cause slipping. At times the friction wheel gets a coating of grease and this also causes slipping. Slipping can be remedied in both cases by setting the machine in motion and holding a piece of old emery paper against the friction wheel and cone.

The pull-cord on all machines should hang eighteen inches from the floor, so that a shearer may easily switch on and off. The cord should not have a wooden knob or a ball of wool attached to the end, for this makes it act like a pendulum, winding the cord around the down-tube. A neat spliced end is the most practical.

Always see that the end of the short down-tube is clean of rust and dirt, so that it will not bind on the handpiece ferrule.

Speed of Machines

The speed at which machines should be run is strongly debated in shearing circles, for experienced faster men like a bit of pace so that they can drive, while learners and slower men prefer the pace to be a bit steadier.

Electric plants are driven at their motor pace, which is usually right. Shafting plants, however, can usually be adjusted to any pace, especially if driven by an engine. The correct shearing pace is to have the shaft running from 650 r.p.m. to 700 r.p.m., and within these limits will be found a pace that should make all shearers happy.

If the plant is too slow the gear will chew and burr when driven hard. If it is driven too fast the cutter goes so fast across the comb that it is like a solid block of metal. It does not cut as well as at a slightly slower pace and produces the effect known to shearers as 'stonewalling'. Excessive pace will also cause the handpiece to run hot.

I remember that in one big shed the shafting was driven by a reliable old engine, its only fault being that it knocked when running at the right pace. If the throttle was turned down a bit the knocking would cease. The throttle had two file marks —the boss's and ours. When he came in and heard her knocking he would smartly turn the throttle back, but it was the job of the 'drummer' down at the end of the board to turn the throttle up again as soon as he went out. When the shearers heard that 'knock, knock, knock', the sheep went out a bit smarter.

It is a good practice to run machines for a while at odd times in the off-season. This helps to ensure that bearings, guts, and working parts are freshly lubricated.

Many shafting plants have done a life-time service, and although really worn out in many parts are still used year after year. All shearing-machinery firms can give a good service in overhauling and modernizing such plants, by using the existing shafting and friction wheels, and renewing bearings and over-head gear. The cost of doing this works out at much less than buying new plant, and the old machines, when overhauled, are ready for another twenty or thirty years of work.

Electric machines with individual motors are particularly handy, because it is possible to unbolt one unit from the wall and plug it in elsewhere. This can be very handy where electricity is available, for dagging out in the yards, the machine being put up in a covered-in dagging pen, thus saving the labour of putting the sheep over the board. The most ingenious procedure I have seen in this connection comprised an electric unit with a good long lead, and attached to rollers which slid along on a pipe running about twenty feet through the middle of an oblong yard. This method of dagging saved dragging sheep up to the machine, the sheep that needed only light dagging being done standing up.

For use in areas that have no electricity most firms now have on the market a simple, compact, and light portable unit that is easily handled and that is excellent when any work, such as dagging or crutching, has to be done outside the woolshed.

CHAPTER NINE

Shearing Competitions

SHEARING COMPETITIONS have gained in popularity of recent years, having a considerable appeal for shearers and also providing a very interesting, action-packed competition for the public. There should, however, be some uniformity in competition rules, as the rules followed in competitions differ greatly at times.

1 There should always be sections to cover the full shearing personnel, in at least three classes—learners, intermediates, and open. The first two classes can easily be restricted to men who have not previously shorn above certain defined maximum tallies. Officials must rely on the honesty of entrants, that they have not previously shorn a tally exceeding the limit set for the class in which they are competing.

2 The most difficult part of shearing competitions is to have both quality of work and speed. It is always essential that the standard of work should be high—perhaps even more necessary than usual when large numbers of people are watching.

Quality of work should be encouraged by the penalties set out below. In a competition, a competitor usually shears only three or five sheep. If his work is marred by serious faults on such a small number of sheep, how many more such faults would appear in his day's work in a shed!

A competitor should be penalised for any of the following faults:

(1) On a ewe, if he cuts a teat to impair the breeding qualities.

(2) On a wether, if he seriously damages or cuts off the pizzle.

(3) If he inflicts a cut on a sheep that causes bad or constant bleeding requiring surgical attention.

Penalty points for the above faults to be reckoned by the number of sheep in the competition divided into 100, i.e. 3 sheep—penalty 33⅓ points; 5 sheep—penalty 20 points; 10 sheep—penalty 10 points; 20 sheep—penalty 5 points. If more than one sheep damaged points added accordingly.

The following rules should also be observed:

(a) If a competitor is a member of a team and inflicts any of the above injuries, 5% points to be deducted for each such injury. (The team is not disqualified.)

(b) On all maiden female sheep a small amount of wool should be cleared above the teats.

(c) On all male sheep a small amount of wool should be cleared around the pizzle.

3 The sheep should be as even as it is possible to get them. The best way to even them up is to take out all the very good-shearing sheep and then the bad-shearing sheep, leaving a middle even line. All sheep used should be lightly dagged.

4 Competitors should draw for stands, but competitors should first be allowed to pass all pens used in their respective heats, and any pen they unanimously agree is not up to standard should be rejected. It is important that this be done before stands and pens are drawn for.

5 Shearers should be allowed one spare handpiece. Handpiece and gear should be the shearer's responsibility. No time allowance can be made for stoppages or hold-ups caused either by handpiece or gear. No time allowance can be made if a sheep gets away on the board, as this also is the shearer's responsibility. However, should there be any hold-up or stoppage through the fault of the shearing plant down to the handpiece, time should be taken for that stoppage and allowed.

6 Every shearer must switch off his handpiece on completing each sheep, catch his own sheep without assistance, and without assistance put the sheep when shorn through the porthole opening.

7 Each competitor should be allowed one man in his catching pen to line the sheep up and generally to act as a second. This man must not touch the sheep when a shearer is catching and must not assist him in any way with switching on or off. It should be watched that there are adequate 'fleece-os' to keep the board clear of wool.

8 There must be a time-keeper for each stand used in the heat, each shearer's time taken from 'Go!' until he switches off at the finish of the last sheep. All shearers should be standing on the board with one hand on the door waiting for 'Go'. The time should be taken by a stop-watch.

9 The judge should inspect each shearer's sheep individually in the pen when shorn. The competitor who has shorn the sheep should tip each one up, since it is underneath that good or rough work stands out.

10 There should be a blackboard or signboard in a conspicuous place clearly setting out each competitor's name and his points when judging is completed, so that both the public and the competitors may follow the judging.

11 There should be a clear indication from the judge at each competition whether socks on the legs are to be taken off or left on. The ruling should be definite one way or the other— either all on or all off. If a competitor leaves socks on when they are required to be taken off, he should be penalised half a point for each sock, i.e. 4 socks left on a sheep means a loss of 2 points.

12 For open or championship competitions each shearer should, if possible, be required to shear five sheep in the competition. In all competitions one trial sheep should be allowed each competitor, the competitor having the right to choose this trial sheep from his pen.

13 Competitors in all classes must gain 75 per cent of quality points (i.e. board and job points) before they can qualify for semi-finals, finals, or places in the final.

Together with the above standard conditions the following methods of judging and assessing points should be adopted:

INTERMEDIATE AND OPEN CHAMPIONSHIP COMPETITIONS

Allowing a maximum of 100 points, this total is made up of three separate divisions:

Time	50 points
Quality of work on the board ..	25 points
Finished job of sheep in the pen	25 points

Taking each of these quotas in turn, points are given or assessed in the following ways:

Time (maximum 50 points)

When the competition is completed the shearer with the fastest time is given the maximum of 50 points. Time points for the other competitors are then worked out as deductions from 50, the deductions being based on either of two tables which we will call Rule A and Rule B. According to these the competitor loses 1 point or part thereof for the number of seconds that his time by the stop-watch is slower than the fastest time.

RULE A: 1 point = 2 seconds per sheep for each of the set number of sheep shorn in the competition.

RULE B: 1 point = 3 seconds per sheep for each of the set number of sheep shorn in the competition.

i.e.,	Rule A	Rule B
1 sheep competition	1 pt = 2 sec	1 pt = 3 sec
3 sheep competition	1 pt = 6 sec	1 pt = 9 sec
5 sheep competition	1 pt = 10 sec	1 pt = 15 sec
10 sheep competition	1 pt = 20 sec	1 pt = 30 sec
20 sheep competition	1 pt = 40 sec	1 pt = 60 sec

Illustrations:

(1) Five sheep competition, Rule A. Fastest time, 6 min = 50 points. Other times, 6 min 15 sec = $48\frac{1}{2}$ pts; 6 min 26 sec = 47.4 pts, etc.

(2) Three sheep competition, Rule B. Fastest time, 5 min = 50 points. Other times, 5 min 21 sec = $47\frac{2}{3}$ pts; 5 min 27 sec = 47 pts, etc.

Deductions can be calculated to the nearest $\frac{1}{3}$ or $\frac{1}{2}$ point.

In any competition where it is required to give more emphasis to quality work Rule B should be used. This table is also recommended for all intermediate competitions and for open competitions where not more than five sheep are shorn by each competitor.

Quality of Work on the Board (maximum 25 points)

Starting with the maximum, points are deducted if second cuts are made, if the belly wool and neck wool are not broken out, and if the shearer loses control of the sheep on the board, thereby upsetting the fleece.

(NOTES FOR JUDGES: It will be found on five sheep that 25 points do not give enough latitude for marking penalties. On 5 sheep, therefore, penalty marks should be $\frac{1}{2}$ points; on 10 sheep, $\frac{1}{4}$ points; on 20 sheep, $\frac{1}{8}$ points, etc. Total penalty marks are brought back by division to an aggregate deducted from 25.)

Finished Job of the Sheep (maximum 25 points)

Sheep are judged for cuts, lumps of two cuts and tassels left on, and for their general appearance. They are judged as a pen and the mark of merit given out of 25 points.

In intermediate and open competitions a final should be run off wherever possible among the top competitors, since this allows the judge to see the best men in action together. The following system should apply to finals:

If 2 stands, 2 or 4 shearers in final;

if 3 stands, 3 or 6 shearers in final; etc.

There should be one judge for each stand used in a competition, but judges should change stands during the heats so that as far as possible competitors' work will be judged evenly by the different judges. When changing stands judges must leave their judging sheets behind them, and if possible use different-coloured pencils.

JUNIOR OR LEARNER COMPETITIONS

These should be judged on style, workmanship, handling, and

the position of the sheep; also on second cuts. Time should not enter into these competitions, though a time limit should be set of five minutes a sheep.

POINTS FOR THE ATTENTION OF JUDGES

(N.B. These points should be studied in conjunction with the Specimen Work Sheet and Specimen Master Sheet, both completed as for an actual competition and showing the use of Rule A and Rule B in different events—see pp. 127 and 128.)

1 See that the competitors' times and their job and board points are transferred from the Work Sheet to the Master Sheet at the conclusion of each heat.

2 Make sure the Master Sheet is entered up clearly *without mistakes*. It gives a bad impression for a Master Sheet to show altered figures, since this sheet is open to inspection by competitors and the public when the competition is completed.

3 Where more than one stand is used, with one judge to each stand, have the announcer change judges on the fastest man so that all judges cover each stand in the competition. On quitting each stand leave your Work Sheet behind so the next judge can continue marking penalties. (Different-coloured pencils avoid confusion.)

4 Board penalties count as marked by each judge and are not averaged. For job points, on the other hand, each judge gives his total and the average is taken.

5 A small allowance may be made for a difficult sheep. However, as the shearer has had the right of a trial sheep the judge should not be greatly influenced by a difficult sheep.

6 Before the competition starts give a clear indication on socks and how points are to be allocated.

7 It is advisable to have a committee member on hand to help work out points at the finish of a competition. There is often a good deal of work involved and it should be done as quickly as possible.

8 Whenever possible hold a final between the top competitors.

9 Judges who have not had much competition experience should practice in the woolshed when shearing is under way.

10 It is essential to keep the same standard all day long. (It will be found that a high standard can be set to begin with, but as the day wears on judges may grow weary and the standard may fall off considerably.)

11 In big competitions it is recommended to have inside and outside judges, i.e. one panel judging the board, another judging the job. This not only saves time but makes things easier for individual judges.

SPECIMEN WORK SHEET

NAME	BOARD PENALTIES Strokes $\frac{1}{2}$ points	BOARD POINTS (out of 25)	JOB POINTS (out of 25)	TIME
A. Dropper	/ / / / / / / / / / / / / / / / $21 = 10\frac{1}{2}$ / / / /	$14\frac{1}{2}$	6 points off 19	11:4
D. Pen	/ / / / / / / / / / / / $18 = 9$ / / / / / /	16	12 points off 13	12:17
T. Smoko	/ / / / / / / / / / $12 = 6$ / /	19	3 points off 22	11:0
N. Fall	/ / / / / / / / / / / / / / / / / / / $34 = 17$ / / / / / / / / / /	8	13 points off 12	10:6

SPECIMEN MASTER SHEET

EVENT: N.Z. Royal Championship
          (Auckland Royal Show)
NO. OF SHEEP (heat):   1 Trial   5 Competition
NO. OF SHEEP (final):   1   ,,   10   ,,
TIME: Rule A   Heats   10 sec. a point
               Final   20   ,,   ,,   ,,
       (Worked out to $\frac{1}{10}$ point)
Heats 5 sheep

ENTRY NO.	COMPETITOR	BOARD PTS. (25)	JOB PTS. (25)	TIME PTS. (50)	TIME	TOTAL
1	A. Dropper	$14\frac{1}{2}$	19	44.2	11:4	77.7
2	D. Pen	16	13	36.9	12:17	65.9
3	T. Smoko	19	22	44.6	11:0	85.6
4	N. Fall	8	12	50	10:6	70
5	S. Grinder	$13\frac{1}{2}$	14	46.6	10:40	74.1
6	L. Emery	17	16	47.7	10:29	80.7
7	D. Comb	6	17	42.5	11:21	65.5
8	Y. Cutter	11	10	49	10:16	70

EVENT: Manawatu Open
NO. OF SHEEP: 1 Trial   5 Competition
TIME: Rule B   15 sec. a point
              (nearest $\frac{1}{3}$ point)

ENTRY NO.	COMPETITOR	BOARD PTS. (25)	JOB PTS. (25)	TIME PTS.	TIME	TOTAL
1	A. Dropper	$14\frac{1}{2}$	19	46	11:4	$79\frac{1}{2}$
2	D. Pen	16	13	$41\frac{1}{3}$	12:17	$70\frac{1}{3}$
3	T. Smoko	19	22	$46\frac{1}{3}$	11:0	$87\frac{1}{3}$
4	N. Fall	8	12	50	10:6	70
5	S. Grinder	$13\frac{1}{2}$	14	$47\frac{2}{3}$	10:40	75
6	L. Emery	17	16	$48\frac{1}{2}$	10:29	$81\frac{1}{2}$
7	D. Comb	6	17	45	11:21	68
8	Y. Cutter	11	10	$49\frac{1}{3}$	10:16	$70\frac{1}{3}$

POSSIBLE INTERNATIONAL SHEARING COMPETITIONS

I have often been asked whether there will ever be an international competition, a World Championship among shearers of different countries. All sheep countries have their own top men who stand out in the art of shearing sheep. It has long been said that a champion shearer is born and not made, and when one considers the thousands of strong fit men who shear and only a numerical few who reach the very top, it could perhaps be right. In view of this there can be little between the top men in each country and tallies and performances will differ mainly as sheep and conditions differ. Because of these differences, the local man in whatever country the competition was being held would have a great advantage, and could be well expected to win, as he would be shearing the sort of sheep he had shorn all his shearing life.

If there was ever a world shearing competition there should be two distinct championships, one for Merino sheep and one for crossbred sheep, as these two kinds of sheep differ so much in the style of shearing required, and together form the bulk of the world sheep population.

Any international rules would have to allow for a competition that covered an eight-hour or nine-hour shearing day. Each shearer's wool would need to be kept separate and fully tested for second cuts, and the shorn sheep would need to be judged by a panel of judges including representatives of each country. Although such conditions would involve so much organization that a competition based on them might not be possible, they would nevertheless be the only fair basis on which a world shearing competition could be held.

The biggest and best-run competition I have shorn in during the fifteen years I have participated in shearing competitions in this country was the Taihape Shearing Championship, shorn under shed conditions in December 1951. First prize money was £100 and a fifty-guinea cup outright. There were thirty competitors, all of them capable of a tally of 300 sheep in a day. A mob of 800 Romney wet ewes was drafted, taking out the 200 best and the 200 worst, which left 400 ewes, clipping an average of nine and a half pounds of wool, there being little to choose between them. There were actually two competitions in one, in that the thirty competitors shore off in heats of five, shearing one trial sheep and five competition sheep. These shearers were all judged, and the first five out of the thirty qualified for the final, those qualifying not necessarily being the winners of the heats. In the final these five top men shore one trial sheep and ten competition sheep. It was a great go. I had a good day, winning the qualifying round and then going on to win the final, shearing the ten sheep in thirteen minutes forty seconds. Competition was held in true shed conditions, the five-stand chute shed being conducted in the same way as for its usual shearing activities.

CHAPTER TEN

Design and Planning of Woolsheds

THE DESIGN and planning of woolsheds is relevant to this text, for the design and layout of the shed can have a considerable effect on the efficiency and comfort of the men who work in it. Too often we find that a shed has been wrongly planned at the outset, causing inconvenience throughout the years, when at no extra cost it could have been planned correctly and helped produce smooth, easy working.

Many different designs of woolshed are in use, ranging from those used for small flocks, and having only one stand, to those used with large flocks and having many stands. Factors to be considered in planning a shed are not only the number of sheep to be shorn, but whether the ground the shed is to be built on is flat or sloping, the best natural lead for the sheep to come in by, and convenience of getting out the baled wool.

For a small standard board shed, I have set out the plan of a three-stand shed (*Fig. 46*) which should meet the needs of a large cross-section of sheep farmers. This shed can be made smaller, or can be increased in size by using up to ten stands on one board. It can also have several wings. Even a very large shed can be planned to use the same basic principles.

There is also set out a four-stand chute shed (*Fig. 48*) which is very practical for a large woolshed, or using several wings into a central board.

These two plans illustrate the two main types of shed—the board shed and the chute shed. The advantages and disadvantages of the two types are:

A. DISADVANTAGES OF A BOARD SHED COMPARED WITH A CHUTE SHED

1 A board shed with a long board involves bringing the wool much further to the wool table than in a chute shed.

2 A board shed congests the wool table more than a chute shed, for the wool is all coming out from one end of the board, so that it is often necessary to have the table on one side of the wool floor. In a chute shed the wool table is in the middle.

3 The active part of the shed—the shearing board—is closed in more in a board shed than in a chute shed, making the board a hotter place to work on, and also making it much more difficult to keep the board swept and clean of wool.

B. ADVANTAGES OF A BOARD SHED COMPARED WITH A CHUTE SHED

1 A board shed can be built on the ground floor, whereas a chute shed has to be raised to allow underneath working of sheep.

2 In a chute shed, if the board is not kept clean, a lot of wool goes down the chutes. In a board shed there is not this danger.

3 It is hard work for a shearer to put sheep down a chute all day. They will not walk out of their own accord as in a board shed.

4 In a chute shed a shearer does not have to cross the board or go so far for his sheep, for it is caught right behind the machine. However, the sheep when caught does not drop into its natural lie for the starting position in the same way as it does when brought across the board to the machine in a board shed. Consequently what is gained in the distance the sheep has to be brought is lost by the effort entailed in getting the sheep in correct starting position.

These are the main differences, and readers can judge the relative values of the two types of shed. As a shearer I prefer a board shed of the type and design set out in this text. My main reasons for this preference are that in such a board shed the sheep go out much more easily, thereby saving physical effort, and also that it is possible to get sheep more easily into the correct starting position, which is so essential for easy and good shearing. However, my preference may arise from the fact that I have shorn mostly on board sheds, and for this reason I would prefer not to be dogmatic in my views on this matter. From the farmer's point of view, the merits of the two types of shed are about fifty-fifty.

The special merits of the board shed plan given here (*Fig. 46*) are that the shearers are shearing directly under the highest part of the roof, in a place where, if good ventilators are put in, there will usually be a cool atmosphere. Board sheds in which the shearers work on the outside, with the roof just above the shafting, make for very hot working. Also in the type of shed shown there is no draught from the portholes, whereas with portholes on the outside wall the board can be very draughty and affect the shearers' backs.

With no outside light coming in from the portholes the sheep will not run out of the catching pens, and this makes possible the use of canvas curtains used instead of a pen door—an arrangement which is a great help to shearers. Sheep also have a tendency to sit more quietly in this central board.

Another point is that the wool comes out in the centre of the wool floor, thereby facilitating wool handling. In wet weather, the sheep do not go straight out into the cold and rain, but, if desired, a complete shed of sheep can be kept under cover.

The internal board allows two entrances for sheep coming into the shed—one in each corner. This is very handy for working separate mobs.

Although the tally pens are under cover, they can be used as a night pen. No actual night-pen space is lost by this, for, when the catching pens are filled and the sheep are packed up a bit in the morning, there is room in the shed to empty the tally pens. Except in dire circumstances it is not good practice to fill the catching pens for use as night pens, as they will get dirty, seriously affecting the wool and making unpleasant work for the shearers.

When working out the sheep capacity of a shed's night pen allow four square feet for each sheep. This of course is for aver-

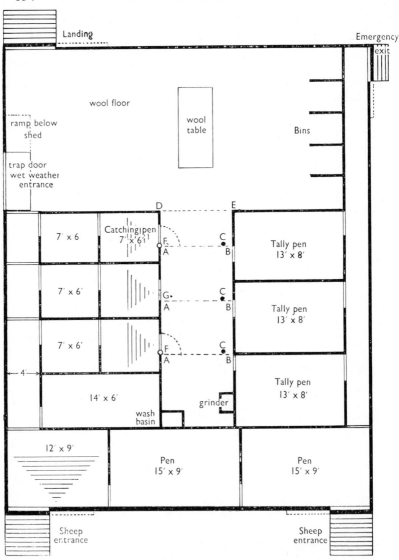

FIG. 46. *Plan for a 3-stand board shed. Windows are not shown, but it should be noted that apart from side windows there should be a set of skylights above the board, each one six feet forward of the individual shearing position (see text). If the shed is built on sloping ground, valuable extra night-pen space can be provided under the wool-floor. Such a night-pen should have a grating floor. A wet-weather ramp from the wool floor down to this night-pen is shown in the plan. The normal night-pens hold from 280 to 300 sheep: extra space below the shed would increase this by about 220.*

age big sheep. It is not good practice to have any individual pen
in the night pen larger than 200 square feet. For tally pens the
sheep, having been shorn, take up much less room, so that a
pen of 150 square feet is usually large enough to take a run of
shorn sheep. A tally pen of this size is often not practical when
the pens are inside, as suggested in this plan, and the usual
practice is to have two count-outs in each run.

Board Shed Plan

Points to watch in the plan of a board shed (*Fig. 46*) are:
. The alignment between A and B is essential, as it gives the
correct position for a catching-pen door in relation to a porthole in order to give a shearer the right shearing position on
the board without having to bring the sheep forward. Note that
the catching door is on the one side of the line A-B, and the
porthole on the other.

The dropper of the machine (C) is set forward of the porthole four inches.

The width of the board (D-E) in this three-stand shed is
7 feet 3 inches, which is all right for up to four stands. If there
are more stands than this the board should be made 7 feet
6 inches wide, as there are more 'fleece-os', each requiring room
to work in. The catching-pen door (F) hangs so that it opens
from the shearer's back: opening this way it will not hit on
the sheep's legs as the sheep is coming out. The smooth part
of this door is put inside (against the shearer's back as he brings
a sheep out) and the rough cleated side out onto the board.
If, as suggested, curtains are used, this door should be made
so that it can fold right back along the board wall, only being
shut between runs. At G are shown canvas curtains—two curtains fixed at the top and down each side, and put up so that
they hang true, meeting in the middle, or slightly overlapping.
The bottom edge is stitched to form a hem, this hem being lightly
filled with sand to keep the curtains hanging straight.

Note the two entrance doors for sheep and also the emer-
gency exit where sheep can be tallied out if the shed is full.

Note that there is a skylight over each shearer. These are
put in six feet in front of each shearer (as he faces the wool

floor), so that he is not shearing in his shadow. Practical blinds should be fitted to these skylights, to be drawn when the sun is shining directly through on to the shearer. Ventilators should be used in the apex of the roof to let hot air escape.

Building paper should be used under the iron above the shearing board and the wool floor as the sheep's breath at night will condense on the iron in the morning, and wet the floor. The grating in the catching pen is laid running across in the same direction as the shearing board, as this arrangement gives the sheep a better grip with the hind feet when catching. If the grating battens are laid running towards the board a lot of sheep will slip down. This cross-ways laying of the grating is sometimes objected to on the ground that sheep's feet catch in the grating, it being argued that the caught foot will pull out of joint. But if the grating is kept in reasonably good order, with the right spacing of five-eighths of an inch between, this very seldom happens. Gratings should also run cross-ways to entrance doors and gates, for this prevents the light from shining through the grating and blocking the sheep. To achieve this arrangement it often happens that the grating in the catching pen has to run opposite to the rest of the shed grating.

The size of the catching pens is important, and should be left as in the plan. The other pens and the wool floor, however, can be made to the sheepowner's desired size or requirements. This plan can be built up on sloping ground to allow for sheep to be held under the wool floor, thereby giving an extra night pen. Note inside loading ramp.

Fig. 47. *Cross-section of floor battens.*
Battens for sheep grating should be machine-dressed and in cross-section be 2 inches deep, $1\frac{3}{4}$ inches to 2 inches wide at the top and $1\frac{1}{4}$ inches to $1\frac{1}{2}$ inches at the bottom. The width of the gap between the battens is $\frac{5}{8}$ inch. In the woolshed plan, patches of broken lines mark the direction in which these grating battens run.

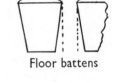
Floor battens

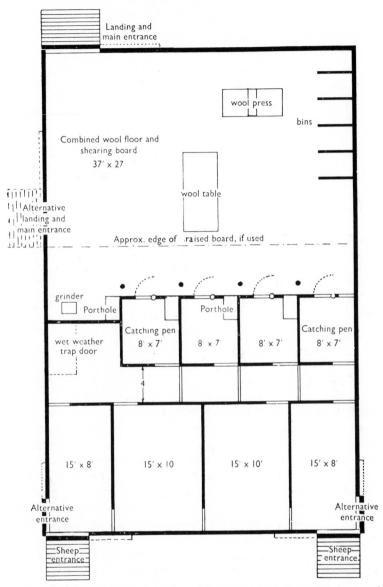

FIG. 48. *General plan of a 4-stand chute woolshed. The normal night holding capacity is from 220 to 250 sheep, but extra night-pen space should be provided under the wool floor to hold an additional 240 sheep. The plan shows a wet-weather trap-door to give access to this holding space. Although four suggested sheep entrances are marked, only two would normally be built. Alternative entrance landings are also marked.*

Chute Shed Plan (Fig 48)

This shed is much the same as the other in main principles. The important points (*Fig. 49*) are:

The distance A-B from porthole to catching-pen door must be approximately two feet, or at least eighteen inches from the dropper to the door opening. With the catching door any closer to the porthole the shearer is likely to kick his handpiece when backing out.

The porthole and the chute can protrude into the shearing board two or three inches, and must have no ledge or lip, but be flush and level with the shearing board, as the sheep's feet must be able to slide across the board down the chute without hindrance.

Curtains can be used for catching pen openings in this type of shed just as in the board shed plan.

It is essential that the base of the shed should be covered in, for if it is left open the wind will blow up the chutes onto the shearing board, making conditions unpleasant. Skylights and ventilators are the same as before. In many chute sheds a raised board is used, i.e. the shearing board is two or three feet higher than the wool floor. If the board is made higher in this way it should be only as wide as a shearer can comfortably shear on, or approximately 5 feet 6 inches wide, for if it is made too wide it will be too difficult for a man standing on the wool floor to reach in and keep the wool clear.

Recently I have seen two devices that save the effort of putting sheep down a chute, and also obviate the danger of wool going down. The first is to have a narrow race leading back into the shed grating instead of a chute. This can be done only in a reasonably small shed as in a big shed it upsets the practical working of woolly sheep. The second is to have the sheep walk out along a narrow short race with a decoy at the far end. This race is pivoted on an axle, so that when the sheep's weight comes over the balance a section of the race tips and the sheep then slides down the chute. These are only suggestions, and I have not had enough practical experience with either to have tested their merits. On a two-stand or three-stand shed a narrow race, leading back (replacing a chute), works very well.

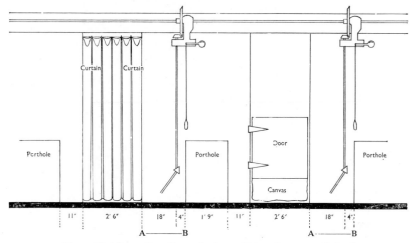

FIG. 49. *Details of porthole wall in chute woolshed*

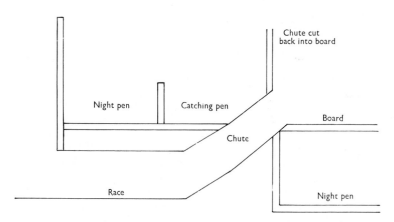

FIG. 50. *Section of chute woolshed*

Safety in the Woolshed

Over the years there have been numerous accidents in wool-sheds. Many factors re machinery and design of sheds contribute to these accidents as well as carelessness on the part of the users. The following is a commentary on these factors, based on my experience and observation over many years in sheds throughout New Zealand.

MACHINERY

1 It is imperative to make sure that the handpiece has the correct cutting tension adjusted before being engaged. If too light a tension, the cutter will fly off and the fork will lock in the comb teeth, and the locked handpiece can travel in any direction. The danger, of course, is that the comb could spear into any part of the body. (For correct adjusting of tension see p. 93.)

2 The leather guard covering the elbow on the down-tube should be kept tight and in good order, as a loose, frayed or broken guard can slip down and expose cogs to wool. Again, the machine will lock and the down-tube can fly around in any direction.

3 It is a dangerous policy to oil the machine in motion, especially on shafting plants, as any frayed apparel can wind around a shaft or pulley. (There has been a recent serious accident through this.)

4 Leather driving cones are much preferred to rubber ones, as the rubber will lift in a layer, the machine will not disengage and frayed rubber has been known to lift completely off and fly at some speed through the shed.

5 On an engine-driven plant, the engine is very often put in a restricted space inside the entrance door with the flywheel, or moving parts on the engine, not guarded. It is dangerous to all people entering the shed. Engines should be guarded when installed and any shafting that is close to the floor should also be housed.

6 Stands should be no less than 5 ft 6 in apart (preferably 6 ft). If any closer it is possible for one shearer to fall back or forward on his mate, when a sheep suddenly puts him off

balance. If the plants are too close together the door of the catching pen can come back with a solid bang and hit the shearer on the head, especially in the long-blow position.

7 Disinfectant should be used in all water pots containing used gear.

GRINDERS

1 All grinders should have a protective guard around the disc, placed so as not to interfere when grinding. (In many cases they do interfere, and for this reason are taken off.) If no guard is used, the disc, travelling at 2,400 revs, has only to be touched to cause a very serious injury.

2 Always see that nuts are tight on the disc. They are designed to tighten as the disc spins, but have many times been known to come off and make a very hurried exit out of the shed. (I have seen one go straight through a corrugated-iron wall.) Through incorrect wiring, or in a spindle-driven grinder belted the wrong way, the grinder can turn in the opposite direction to what it should, and of course wind the nuts off.

3 The grinder should always be installed out in the open with plenty of light. In a restricted place with poor lighting accidents easily occur.

4 Always ensure that the holder for the grinder has a good strong magnet. (Most holders in sheds need remagnetising.) Also that the holder pins are long enough to hold the cutter when it tips back at the top. If the magnet is weak and the pins short, the cutter can easily fly off when it touches the disc. It will always fly up, and this has caused a lot of injury. (For setting and adjusting of grinders, see p. 85.)

SHED

1 See that the steps into the shed are in good repair, especially sheds which are high off the ground. A leg is easily broken if a step gives way.

2 Doors of catching pens should be low so that a shearer can easily see over them and the sheep in the pen can see the approach of the shearer. There have been at least two fatal accidents where the last sheep in the pen has charged out into the light when the door was opened, meeting the shearer head on.

3 Weights that hold up sliding gates should be housed. Should anyone be in the way when a weight comes down suddenly, a serious injury could result.

4 A good smooth shearing board is required. Shearing on a broken or badly patched board, the shearer can very quickly get sore feet. Shearing should always be done on a wooden floor.

5 Grating in the catching pen needs to be in good repair (spacing $\frac{5}{8}$ of an inch), as faulty grating can cause injury to sheep, as well as straining the shearer when he brings the sheep out, if a hind foot is caught in the grating.

6 Good lighting is essential, especially on the shearing board, round the grinder, and over the wool table. This is best served by a skylight in the roof. A skylight for a shearer requires to be 6 ft in front of his shearing stand so he is not shearing in shadow. Poor lighting results in eye strain and can also result in damage to both sheep and men.

7 Killing sheep in the shed is not a good practice as there is a big danger of blood poisoning infecting the shed. Once in a shed, it is very difficult to eradicate and can result in a big loss of sheep and also ill-health of workers.

8 Wool presses: Recently there have been several accidents involving wool pressers. The main danger point is on a wind-up press with a centre pole. The winding up is all right, but when winding down the rollers on the pole sometimes stick and the handle is let go by the worker to free the top. When freed it drops with a sudden jolt, resulting in the iron handle spinning around with force and striking the worker, in most cases on the arm or head. Also, on some old presses a worn ratchet is a dangerous part, and really needs building up with metal. If used worn it should be treated with respect, as it is prone to slip when the press gets a good strain.

9 Hobnailed boots should never be worn. On a greasy floor some good spills can easily happen. Rubber or leather-soled footwear is all right, but sack moccasins are much preferred for all shed workers.

10 Hand shears or blades: It is not a good practice to have a number of daggers working in a small pen with blades. I know of several cases where a shearer catching a sheep has been forced back on to blades held by his mate who has just completed a sheep.

11 A good first-aid outfit should be fixed in a place handy to the board, with the telephone number of the nearest doctor painted on the cabinet.

Design of Sheepyards

As NO WOOLSHED is complete without a set of yards, plans of two of the best types of yards I have seen in operation are shown in *Figs. 51* and *52*. *Fig. 51* is an excellent economic yard, and with the required holding paddocks there is no limit to the number of sheep such a yard can handle. *Fig. 52*, the circular yard is, of course, more costly, but where it can be afforded is an excellent yard.

The following are the basic points to watch in planning yards:

1 The lie of the land is, of course, important, and it should be remembered that sheep work and draft best on a slight upgrade. On such a grade the sheep are also more easily seen by the man on the drafting gate when walking up hill. However, it is not a good practice to have yards on a steep slope as there is a danger of sheep smothering in the low corners.

2 Good drainage is a necessity. There are too many poorly-drained yards that quickly become churned into mud, making conditions difficult and unpleasant for working sheep.

3 Shelter and shade are grand things to have in yards, and where conditions are favourable a few willow trees planted near by are greatly appreciated by sheep and man in hot weather. A plantation or break on the windward side of the yard is a good idea, and, as already mentioned, can provide valuable shelter at shearing time.

4 To have water at the yards is always an advantage. Most yards either include a dip or have the dip near by, and a good water supply is essential. In big yards a water trough is an

asset, and it has long been contended that there would be fewer losses at dipping time if sheep were not dipped when thirsty.

In Australia yards are often sprinkled for dust, some places using a permanent sprinkler system. Hot, dry weather at shearing time is experienced in all countries, and a good water supply will quickly combat dust.

5 Size of yards. Yards, of course, vary in size, subject to the number of sheep and number of separate mobs they will be required to handle. In planning the size of yards, allow five square feet per sheep—that is, for grown average to big sheep —which will allow ample room to work sheep around yards. In dry, fine-wool country where sheep are not as big as cross-bred sheep, three and a half square feet per sheep can be reckoned on.

6 In taking into account the lie of the land, and the natural way sheep will run, the drafting race is given first considera-tion and the rest of the yards built around it.

Yards are usually made up of the following:

Receiving yards. These must have the capacity to take incom-ing mobs. They communicate with the *forwarding yards*, or *diamond pens*, which are much smaller, and lead up to the *crush pens* or *forcing pens*. These last are usually one or two long pens that communicate by a two-way gate with the mouth of the *drafting race*, which is long and narrow, just wide enough for the sheep to pass through in single file. This drafting race divides sheep into the separate lots required, and these are then handled by the *drafting pens*, *check pens*, and *holding yards*, which should correspond in size to the receiving yards.

It is not the intention to set out details of these yards, but the following are a few points to be noted about crush pens and the drafting race:

Crush pens. The shape of crush pens should be long and narrow, so that it will be easy for one man to pack sheep up. The ends of crush pens should gradually taper off into the race, and should not narrow abruptly. In a double crush (i.e. two pens) a good long gate, of approximately four feet width, should be used at the end of the dividing fence to lead the sheep from either pen into the race.

The width of crush pens has long been a debatable point. The best plan is to have one crush wider than the other, one 3 feet to 3 feet 6 inches, the other 5 feet wide. With this type, when only one man is working sheep, for drenching, etc., he can use the narrow crush, and if two men are working the wide crush can be used.

The reason for having two crush pens is the time-factor. As soon as one is emptied, the other, already filled, is brought into use, there being no time lost in filling. The sheep in the full pen also act as decoys for the incoming sheep. It should also be noted that where crush pens taper into the race, this tapered part should be close-boarded to prevent sheep seeing the man on the drafting gate.

Drafting race. This is important, for a bad drafting race will spoil the best of yards. The length can be anything from 10 feet to 25 feet, depending on the number of drafting gates used. The height should be approximately three feet, which is low enough to allow a man to reach over and help the sheep along, and yet is high enough to prevent the sheep while passing through from seeing the men and dogs around.

The drafting race should be close-boarded of smooth dressed timber, without any protruding bolt-heads, cleats, etc. The floor of the race is best put in of concrete, for an earth floor soon wears away.

The width of the race can vary according to the type of sheep being drafted, and adjustable races are sometimes used. However, a tapered race, made to the following dimensions, will take sheep from the biggest to the smallest without much difficulty: width at the bottom 11 inches, width at the top 22 inches. This tapered race is as easy and economical to construct as a straight-sided race, will accommodate the largest sheep, and is narrow enough low down to suit the small sheep. The drafting gates must be of reasonable length—approximately 3 feet 6 inches— for if shorter they will form too sharp an angle, and sheep will have difficulty in getting around quickly. The gate should be light, strong and smooth, and if it is made by laminating dressed boards together it will give a dressed smooth edge both sides.

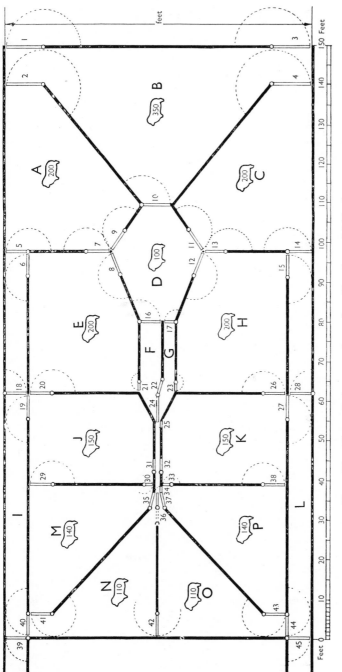

FIG. 51. *Plan of sheepyards to handle about 1000 sheep*

Care should be taken to see that no nails, bolt-heads, etc. are protruding, as sheep can easily be injured here. To stop bruising, especially on lambs, a good safeguard is to pad the leading edges of the gates with rubber from an old tyre. By using the double drafting gate sheep can be drafted three ways at once. In large holdings three gates are sometimes fitted in order to provide for four-way drafting.

The following plans and specifications of two main types of yards are reproduced by permission from *Design and Construction of Sheepyards*, by J. E. Duncan.

Medium Yard (*Fig. 51*)
To handle about 1000 sheep

The main feature of this plan (medium yards to handle about one thousand sheep) is that with the race and gates shown sheep can be drafted six ways. Of course it may be asserted that this is quite unwarranted in a yard of this size, but it is included to show the general arrangement required.

The gates 35, 36, and 37 give four-way drafting and by stationing a second man at the gates 31 and 32 sheep may be drafted off two additional ways into pens J and K. If a simpler set of yards is required, gates 31, 32, and 36 may be omitted, giving normal three-way drafting. In this case the division between pens N and O would be eliminated and they would be joined into one large pen holding 220 sheep. If it is felt that this would be too large, the end of the yard could be moved in a little to give smaller drafting pens.

Another and in some ways a better method of arranging for three-way drafting is to retain gate 36 but to hang it about 3 feet 6 inches further back in the position shown dotted in on the plan. Then the free end of this gate will swing over to the fence on either side (just behind the gateposts of 35 and 37) so that when it is fastened one way sheep may be drafted into pens M, N, and P; and when it is fastened the other way they will go into M, O, and P. Should the sheepowner decide to follow this arrangement in the first instance, it will be advisable to have the gateposts of 35 and 37 somewhat closer together than as shown for four-way drafting.

Another useful feature is the provision of the two alleyways I and L. These serve as right-of-ways connecting the various pens and greatly facilitate the moving of sheep. They may also be used as lead-ins to the woolshed and the dip where this can be arranged conveniently. As an alternative a small concreted section could be closed off by gates and could be used as a foot-rot bath or for feeding a foot-rot trough outside the yards.

The small gates 21, 22, and 23 in the crush pen, though by no means essential, are a great convenience in moving odd sheep around.

ESTIMATE OF QUANTITIES

Fences:	Length of outer fence	..	..	428 feet
	Length of fence round diamond	..		43 feet
	Length of other internal fences	..		533 feet
	Length of all fences (excluding gates)			1004 feet

If five 4 in. x $1\frac{1}{4}$ in. rails are used for fences, a total of 1004 x 5 equals *5020 lineal feet* of this timber will be required.

Crush and Race

End panels of the crush on both sides are close boarded with single-dressed 6 in. x 1 in; equal 17 ft. x 6 equals *102 lineal feet.*

Sides of race are close boarded with single-dressed 6 in. x 1 in.; equals 30 ft. x 6 equals *180 lineal feet.*

If drafting gates and race stop-gates are constructed of double-dressed 6 in. x 1 in. timber, add *120 lineal feet*, plus *60 lineal feet of 4 in. x 1 in.* (dressed) for stiles.

If drafting gates are constructed on the laminated principle (three layers) to give a flush, smooth finish, a total of 120 feet of 6 in. x 7/16th in. finish dressed timber will be required instead of the 42 feet of 6 in. x 1 in. and 24 feet of 4 in. x 1 in.

Gates

If common swing gates are constructed, the following timber will be required:

For four 10 ft., one 8 ft., twenty-six 6 ft., three 4 ft. 6 in., and four 2 ft. gates a total of *2197 lineal feet of 4 in.* x *1 in.* timber for stiles and stays of all gates, and rails of all gates 7 feet and under, plus *240 lineal feet of 4 in.* x *1¼ in.* timber for rails of gates 7 feet or wider.

Total timber (lineal feet): 5260 of 4 in. x 1¼ in., 2257 of 4 in. x 1 in., 402 of 6 in. x 1 in.

Posts

A total of 180 posts will be required and of these at least forty-five require to be heavy enough to act as gateposts.

Circular Yard (*Fig. 52*)
To handle 2000 sheep or more

There are apparently many who assume that circular yards can be built of practically any size, but a little reflection will show that this is not possible with the usual type shown here. There must be room for the crush, the race, and the drafting pens, not to mention part of the diamond, between the centre and the outside fence of the yards. This imposes a certain minimum size which is not much below that shown in this plan.

Though the plan is largely self-explanatory, one or two features can be elaborated on. The central diamond pen (I) can be used to great advantage in circular yards and is really an essential part of the design. In this case it communicates with not fewer than ten different pens. The outside race or alleyway also serves as a connecting link between the majority of the pens. In the plan only two entrances to the yards from outside paddocks are shown, through gates 2 and 13, but of course fences and gates can be included at any convenient point round the outside ring fence to suit existing conditions.

Four-way drafting is a feature of the design, and circular yards lend themselves to this. The arrangement of gates leading to the woolshed should also be noted. The end of the ramp is

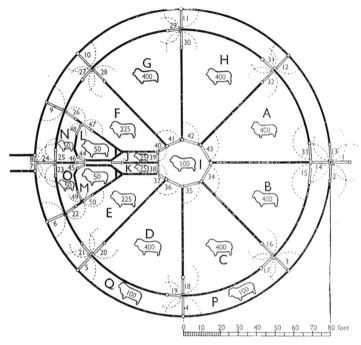

FIG. 52. *Plan of sheepyards to handle 2000 sheep or more*

closed by two small gates (7 and 8) each four feet long. They may be swung across to meet gates 23 and 25 of the drafting pens so that sheep from pen N or O can be run through directly into the shed; also by swinging gate 24 to meet gate 7 or 8, sheep may be run into the shed from either half of the outer ring of the yards.

ESTIMATE OF QUANTITIES

Fences:	Length of outer ring fence	..	475 feet
	Length of inner ring fence	..	360 feet
	Length of other internal fences	..	546 feet
	Length of all fences (excluding gates)		1381 feet

If five 4 in. x 1¼ in. rails are used for fences, a total of 1381 x 5 equals *6905 lineal feet* of this timber will be required.

Crush and Race

End panels of the crush on both sides are close boarded with single-dressed 6 in. x 1 in.; equals 16 ft. x 6 equals *96 lineal feet.*

Sides of race are close boarded with single-dressed 6 in. x 1 in.; equals 28 ft. x 6 equals *168 lineal feet.*

If drafting gates and race stop-gates are constructed of double-dressed 6 in. x 1 in. timber, add *84 lineal feet* plus *48 lineal feet of 4 in. x 1 in.* (dressed) for stiles.

Gates

If common swing gates are constructed, the following timber will be required:

For nineteen 10 ft., two 9 ft., twelve 8 ft., nine 6 ft., four 5 ft., and two 4 ft. gates a total of *2379 lineal feet of 4 in. x 1 in.* timber for stiles and stays of all gates, and rails of gates under 7 ft., plus *1520 lineal feet of 4 in. x 1¼ in.* timber for rails of gates 7 ft. or wider.

Posts

A total of 226 posts will be required, and of these at least fifty-three require to be heavy enough to act as gateposts.

CHAPTER TWELVE

Sheep Breeds and their Shearing Qualities

AS THE SHEARER'S work brings him in contact with many different breeds of sheep, the following notes about the popular breeds and their shearing qualities will, I hope, be found helpful. I have had practical experience of shearing all the breeds mentioned here, and the commentary is based upon that experience. It describes conditions with average-shearing sheep carrying twelve months' wool. It has to be remembered, of course, that local and seasonal conditions affect the shearing quality of sheep, and that there are good-shearing and bad-shearing sheep in all breeds.

Merino

The Merino was the foundation sheep of the New Zealand wool industry. The breed appears to have originated in Spain and Morocco, but the development which has produced the modern Merino took place in Australia, and the Merinos used in this country are of Australian origin. Although not so widely used now as formerly, the Merino is still the main breed on the really high country of the South Island, where there are still large numbers of them, mainly in Canterbury, Marlborough and Central Otago. It is the only sheep that stands up to the severe conditions prevailing in the hard snow country, as it is a great climber, lives on a small amount of feed, and can withstand snow conditions better than any other breed. Merinos stand up well to extremes of heat and cold, so long as conditions are dry, but do not thrive under wet conditions. Merino wool is the finest quality wool in the world, and has a wide range, varying from the normal fine wool to superfine quality. The bulk of Merino wool has a count of from 60's to 70's, going up to 90's in superfine flocks. The breed is still widely used to cross with long-woolled breeds to produce the half-bred.

SHEARING QUALITIES

The Merino is probably the most difficult of all sheep to shear, for the following reasons.

1 The large number of wrinkles and folds in the skin, especially around the neck.

2 A large proportion of Merinos have horns, so that shearing the head is more difficult and consequently slower than with other breeds.

3 They have a light, tender skin that is easily cut or torn.

Straight Furrow photo

FIG. 53. *Merino ram*

4 The Merino is run on hard, dry, and low-producing country.

5 Where the Merino is run on mountain country, or shale, and in sandy areas, the fleece often carries a lot of grit and stones, which, in the dense wool, can make shearing really hard.

6 The dense and fine Merino fleece often has a hardened yellow yolk at the base of the wool down near the skin. I have seen some Merinos in which the wool close to the backs of the sheep was as hard as concrete.

However, with the modern trend of breeding, tending to eliminate excessive skin folds and wrinkles, there are many nice-cutting reasonably plain-bodied Merinos that shear well. In addition, the breed has some real shearing advantages. The Merino is a very docile sheep on the board, compared with the coarser, long-woolled breeds, which are always ready to kick and generally to resist the shearer.

Straight Furrow photo

FIG. 54. *Merino ewe*

The fine wool of the Merino is really good, soft cutting, compared with the hard cutting on the stronger-woolled breeds. On a clean-woolled Merino, a comb and cutter will last two or three times as long as on strong-woolled sheep. It is to be remembered that the Merino is not the only breed that gets sand and grit in its wool.

Shearers who have always shorn Merinos like shearing this breed, and seem happier with it than with the long-woolled breeds, one reason for this being that the short staple of wool on the Merino allows a shearer to see and handle his work more easily. With the long wools, quite a lot of hand-work is done 'on the blind'. Ten pounds of wool on coarse, long-woolled types, makes a great bulk compared with a Merino fleece of the same weight. I believe also that there is less physical strain on the shearer with this breed, for Merinos, run on hard, dry country, are much lighter to handle than 'dual-purpose', or mutton breeds run on high-producing lands.

Summed up, my opinion of the Merino is that of all breeds, average to average, it is the slowest and most difficult sheep to shear. However, although a shearer's tally with Merinos may not be as high as with other sheep, the rates of payment are higher than for most other breeds, in some cases in New Zealand being as much as double the standard rate.

Corriedale

The Corriedale is a very popular breed, and now has the second highest numbers in the world sheep population, being exceeded only by the Merino. The Corriedale was evolved in New Zealand by crossing the Merino with the Lincoln, and breeding up the resultant progeny, until after many generations of inter-breeding, the results of the Merino-Lincoln cross, a definite type was established, and the breed became famous as the Corriedale. Like the Merino, the Corriedale does not favour wet climatic conditions, but thrives best on dry hill country. It also does well on flat country providing the rainfall is not too great. The Corriedale has a fine wool, which inherits much of the Merino quality, but which has the distinctive heavy staple of the Lincoln. The Corriedale is a very popular dual-purpose breed, with a good heavy fleece of quality wool, and an excellent carcass, which makes the Corriedale ewe ideal for producing fat lambs. The breed is widely used in Canterbury, Otago, Marlborough, and to a large extent in the back country, but it is better suited by medium hill country. The wool count of the Corriedale generally averages 56's, with a range of 50's to 60's on heavier country, and 56's to 60's on high country.

SHEARING QUALITIES

The average Corriedale is quite fair shearing, and while it is not a high-tally sheep, it is easier to shear than a Merino, being plain-bodied apart from an odd neck wrinkle or dew-lap, and lacking the horns of the Merino. The Corriedale is well covered with wool, and unlike heavy breeds grows more wool than hair on the hocks, thereby making it imperative that they be shorn trimmed to the feet. These points or socks do not have a tendency to lift or rise off the skin as they do on some breeds, but comb off quite well.

Straight Furrow photo

FIG. 55. *Corriedale ram*

With a reasonably fine, long, heavy staple, the Corriedale produces a big weight of wool, and yet has the tender, loose skin of the Merino. This results in the wool often pulling the skin out, which makes it sometimes difficult to 'drive' with a full comb without marking the sheep.

Straight Furrow photo

FIG. 56. *Corriedale ewe*

Corriedales are docile and handle and lie well on the board. I am of the opinion that they shear out of season—that is, in winter and early spring—better than cross-breds or long-wool breeds, many of which are extremely difficult to shear in these months. For this reason, Corriedales suit pre-lambing shearing. Shearing rates are quite high for Corriedales in New Zealand, and taking this into account, the breed, being fair average shearing, is quite an attractive proposition for the shearer.

The Half-bred

The Half-bred is very similar to the Corriedale in type, and in its uses. The Half-bred, which is of the same ancestry as the Corriedale (Lincoln-Merino, or English Leicester-Merino, mainly), is also virtually confined to the South Island. The wool is similar in quality to that of the Corriedale. Half-breds are second only to Merinos in climbing and foraging ability, and are widely used on the high country. The cull ewes from the stations find a ready market with flat-country farmers to cross with mutton breeds for fat-lamb production. Although the English Leicester-Merino and the Lincoln-Merino are the most widely-used crosses, the Romney-Merino cross is also fairly popular, particularly in Otago. Half-breds are recognized in New Zealand as a cross between Merinos and one of the English longwool breeds.

Straight Furrow photo

FIG. 57. *Half-bred ram*

SHEARING QUALITIES

There appear to be different types of Half-bred sheep—those that throw to the Merino, and those that throw to the longwools. The average Half-bred usually shears reasonably well, and in its shearing qualities is very similar to the Corriedale. It is usually run on high-altitude, low-producing country, which naturally impairs its shearing qualities. My experience of Half-bred shearing is that owing to type and the country they are run on, shearing can greatly differ in this breed. Some Half-breds go very well; some very poorly. But I would say that average to average they are equal shearing with the Corriedale.

On the three finewools (Merino, Corriedale and Half-bred) the points of the comb must be kept quite a bit brighter than for shearing other types of sheep, but the finewool comb still requires points rounding off and polishing on wood, as already described in this text.

The finewool shearer must use a lighter and more flexible hand to cover wrinkles and enter finewool points, etc., than for the flowing follow-through style of the longwool shearer, who shears more of a tight-skinned, plain-bodied sheep.

Any shearer who is used to finewools usually prefers shearing this type of sheep, as the wool is good cutting, the sheep are good handling, the rates of pay are attractive, and the fleeces are not nearly so matted as with the longwool breeds. However, it must always be remembered that the finewool sheep are generally slower to shear, so that it is not easy to make a high tally with them.

The Romney

The Romney is New Zealand's most popular sheep, for it is extensively used in almost every province of the Dominion. Originating in England, in the Romney Marshes of Kent, it is an old-established breed of noted constitution, with ability to stand up to a wet climate. In the North Island, where Romneys comprise the greater part of the total ewe flock, they have had to adapt themselves to a wide variety of conditions, whether the country is flat or hilly.

New Zealand breeders have evolved a different type of sheep from the original Romney imported from England, so that today the typical New Zealand Romney is low-set, with a good meaty carcass, and a greatly improved wool covering. A great demand exists for New Zealand Romneys in other parts of the world, and there is a large export trade in this breed. It is essentially a dual-purpose animal, having a good carcass for mutton, and producing a good weight of fair quality wool of the 'Crossbred' designation, the wool count being generally 44's to 50's. Fat lambs from Romney ewes by rams of the Down breeds have few superiors for the meat trade. The Romney ewe is to the North Island what the Corriedale and the Half-bred are to the central and northern parts of the South Island—the mainstay of the farmer's ewe flock.

Straight Furrow photo

FIG. 58. *Romney ram*

SHEARING QUALITIES

From a shearer's point of view, Romneys differ greatly in quality. I have seen my day's full tally drop from 300 to 150 when moving from a good-shearing flock to a tough-shearing flock. The average Romney in winter and early spring does not shear well: in many of them at these times the wool seems to be almost glued on to the skin, so that it is common to see shearers poking the wool off, taking five or six minutes a sheep. However, there is a notable difference in the shearing qualities of this breed when the yolk rises. From October to January I would say the Romney provides some of the best shearing of all breeds—maybe not quite the fastest, but the most pleasant shearing for a day-to-day tally.

The Romney is a well-woolled sheep, covered right down on to the points. At the right time of the year these points lift or rise off the skin allowing the comb and the handpiece to enter easily and the blows to finish out on the points without poking.

Although the Romney is not as docile on the board as the finewool, it is a reasonably good-handling sheep.

Romneys produce a lot of grease in the wool, so that the wool combs nicely. With its tight skin, open wool, and plain body, the good average-shearing Romney lends itself to consistent, steady, reasonably high-tally work. It must be remembered, however, that the sheep do not shear themselves.

Romneys are big sheep and can become very heavy when run on high-producing areas. They clip heavy wool weights, so that the fleeces can often become very matted, especially on the bellies and necks, and while high tallies can be shorn with them, the work demands a great deal of skill and physical fitness. The few 'gun' shearers who in a day do three hundred big Romneys each averaging ten pounds to twelve pounds of

Straight Furrow photo

FIG. 59. Romney ewe

wool, must be termed masters at their work, as they have not only to cover big, well-woolled sheep, but must catch and handle across the board liveweight approaching twenty tons of mutton a day.

Summed up, the Romneys at the right time of the year are nice-shearing sheep, but at the wrong time, when the yolk is not up, or when the sheep have been 'done hard', they can be among the toughest a man can put a handpiece into.

Cheviot

This Scottish breed, with the characteristic bald head, is becoming increasingly popular with farmers in the North Island, particularly where the country is high and rough with a covering of fern or scrub. The Cheviot is hardy and active, and thrives well under conditions which would test the constitutions of most breeds of sheep. It is filling a gap in the North Island hill country, by being used on country which previously carried

Straight Furrow photo

FIG. 60. *Cheviot ram*

Romneys, but which in reality was not suited to that breed.

The Cheviot's main assets are its hardiness and agility, and its ability to thrive under poor conditions. The breed stands up well to cold, wet conditions, where sheep have to forage for feed, and where less hardy breeds would find it difficult to live. Cheviots have good carcasses, and are exceptionally good mothers. They are widely used today for crossing with Romneys, and this cross is becoming more popular on rough country, for with their 'open' faces, Cheviots are not affected by wool-blindness.

The breed clips well, and the wool is reasonably fine, in the quality range 48's to 54's.

SHEARING QUALITIES

The Cheviot is a sheep that looks such a beauty to shear, and yet in reality is disappointing. As it has bare points, no wool on the head or legs, one sometimes hears said of Cheviots

Straight Furrow photo

FIG. 61. *Cheviot ewe*

that 'there is nothing left for the shearer to do'. This statement is a bit wide of the mark, as a shearer still has to cover the body of the sheep, and if a sheep's points are open and good combing, they take only a few seconds to shear. It is true that the Cheviot is good fast shearing, but it is not so good or so fast as is reckoned by those who have never shorn this breed.

The worst fault of the Cheviot from the shearer's point of view, is its wild, restless, fighting manner when on the board; it is by far the most active sheep of all breeds on the board. Cheviots seldom sit or lie still for shearing, and will continually kick, wriggle, tuck their legs up, and fight the shearer. The last one in the catching pen will most likely jump out; the first one in the tally pen will try to jump out. Recently at Massey College we had one jump out of the pen, and then go over seven wooden yard fences till he hit the paddock.

The other factor that impairs the shearing of this breed is that with an average finewool, it is often very dry, and devoid of grease, which results in dry, 'doughy' cutting. However, the average Cheviot is good fast shearing.

Border Leicester

The Border Leicester is fairly widely used in Canterbury, Otago, and Southland, the Border Leicester ram being popular with many fat-lamb producers because of the heavy weights attained by his progeny as lambs, and their quick-maturing quality. In some districts the breed is crossed with the Southdown to produce Border-Southdown rams for fat-lamb production.

The Border Leicester is descended from the Cheviot, and inherits the same bald head, which obviates wool-blindness. Although primarily a mutton and fat-lamb breed, the Border Leicester has a fair fleece of wool, in the 44's to 46's range.

Shearing Qualities

The Border Leicester is one of the best-shearing breeds, because of its bare points and open free-cutting wool. It is a sheep that is inclined to struggle on the board. It is long-bodied and a bit leggy—qualities which slightly impair its shearing qualities. Taken on average, however, the Border Leicester is a good-shearing sheep.

Straight Furrow photo

FIG. 62. *Border Leicester ram*

Straight Furrow photo

FIG. 63. *Border Leicester ewe and lamb*

English Leicester

The English Leicester is another of the British breeds which has played a notable part in developing New Zealand's sheep farming, but is now more or less on the way out. It has, however, been widely used in the production of Half-breds for South Island conditions, and that is where its main importance still lies. Today there are very few English Leicester ewe flocks, and most of these are in Canterbury. The breed has a good fleece of strong wool of 40's to 44's quality, with a fairly long staple. It crosses well with Down breeds for fat lamb production. The Romney, however, has taken over most of the job once done by the English Leicester.

SHEARING QUALITIES

The English Leicester is much like the Border Leicester for shearing, with its free open-cutting wool, but it differs from the Border in that it has woolly points. The wool on the legs and head, however, is locky and open, and does not retard shearing, in that it combs very well. The English Leicester is closer to the ground than the Border, and is a better-shaped

Straight Furrow photo

FIG. 64. *English Leicester ram*

sheep for shearing. It is a docile sheep on the board, and in its shearing qualities is equal to the Border. Old shearers, remembering the days when there were many English Leicester flocks, often say what grand shearing these sheep were.

Lincoln

Lincoln sheep were much more used in New Zealand in the early days than they are now. They have had a considerable influence on our present-day types of sheep. It is to the crossing of the Lincoln with the Merino that we owe the origin of the Corriedale. Also, the Half-bred as used in the South Island high country is mainly of Lincoln-Merino origin. There are still quite a number of Lincoln stud flocks in New Zealand, but their main importance today is in providing rams for crossing with Merino ewes. The wool, ranging in quality from 36's to 40's, is remarkably heavy, with a long, broad staple, and a very pronounced crimp.

Straight Furrow photo

FIG. 65. *Lincoln ram*

SHEARING QUALITIES

The Lincoln is a big strong-woolled sheep, with plenty of vitality, so that it can be a bit hard to hold on the board. It is just average shearing, although occasional Lincolns can be very good. The wool is strong and hard-cutting, and will test the best of gear. The wool also very easily becomes matted. The Lincoln clips a big weight of wool per sheep, and the heavy long staple is inclined to pull the skin out. This, combined with matting, makes it extremely hard to shear a Lincoln without marking it. Summed up, it can be said that Lincolns are just average shearing.

Matted wool often wears the skin off the backs of the shearer's fingers, and it is common to see the back of the forefinger on the handpiece hand bleeding, with the skin worn through.

Southdown

The Southdown is easily the most popular of the fat-lamb breeds. It is one of the oldest British breeds, and comes from the southern counties of England, having been developed mainly in Sussex and Surrey. Quick-maturing, well covered

Straight Furrow photo

FIG. 66 *Southdown ram*

with meat, and hanging up attractively 'on the hooks', the Southdown has long been regarded as the premier meat breed of New Zealand. Southdown rams are widely used to cross with Romneys, Cheviots, Corriedales, Half-breds, and indeed, with every type of ewe in this country, to produce quality fat lambs. Although not producing a fleece of much weight, the Southdown has the finest wool of any of the British breeds. It is of 58's to 64's quality, but it is short in the staple.

Shearing Qualities

The Southdown is the hardest of all the mutton breeds to shear. On the points, the legs, and the head, it has a great deal of wool that is usually very hard to poke off. Having a squat thickly-set, and strong body, the Southdown is very difficult to bend into correct shearing positions. Further, it has a fighting disposition, and this does not help. As the Southdown is run in high-rainfall areas, being a mutton sheep, it is very common for the wool to develop a hard yellow yolk that sets like concrete on the back of the sheep, which makes for hard shearing.

Straight Furrow photo

Fig. 67. *Southdown ewe*

Southdown rams are particularly hard to shear, and it is a strong man that can bend an agile Southdown ram's neck and head when the sheep makes up its mind to hold itself stiff. I have seen some wild moments when shearers have lost their temper with Southdown rams.

Considering the average, I could not say the stud Southdown is a good-shearing sheep. It is definitely the hardest to shear of all the mutton breeds.

Suffolk

The Suffolk is one of the world's most popular mutton and fat-lamb breeds, and is now being used fairly extensively in this country. The breed originated in England from crossing the Southdown ram with the Norfolk ewe. It has an exceptionally good butcher's carcass, and the lambs mature early to very good weights. This breed is bound to be popular with those wanting heavy-weight lambs. The wool is of down type and averages 54's.

Straight Furrow photo

Fig. 68. *Suffolk ram*

SHEARING QUALITIES

The Suffolk is the best shearing of all the mutton breeds, with a nice-shaped body for shearing, clear points, legs and head, a strong, tough, tight skin, and a fine wool that cuts well. All these factors combine to make the Suffolk a very good shearing sheep.

Straight Furrow photo

FIG. 69. *Suffolk ewe and lamb*

Dorset Horn

The Dorset Horn is remarkable for its ability, if necessary, to produce two crops of lambs a year, although in practice, breeders do not require their Dorset Horn ewes to do so. However, it is an early-maturing breed, and as such is very useful in the fat-lamb industry. Originating in the West Country of England, the Dorset Horn is now used fairly extensively in New Zealand, mainly on the flat country where fat-lamb and mutton production predominate. The wool is reasonably fine and dense with an average staple. The quality ranges from 50's to 56's.

Straight Furrow photo

FIG. 70. *Dorset Horn ram*

SHEARING QUALITIES

The Dorset Horn is a big-bodied mutton-breed sheep, that shears fairly well, being very clean around the points, which are lightly covered with wool. It has a nice-cutting wool.

The Dorset Horn is not the best natured of sheep. The main trouble for the shearer is caused by the horns, which become real obstacles, for the Dorset Horn does not sit quietly like a Merino, and the horns get in the road when a sheep moves about on the board.

In spite of all this, however, the Dorset Horn is a fair average-shearing sheep.

South Suffolk

This is a New Zealand type derived from crossing Southdowns with Suffolks, and is being very widely used in fat-lamb farms in Canterbury and is spreading rapidly to other parts of New Zealand. The South Suffolk has most of the attributes of the Southdown, together with the extra weight gained from the Suffolk. It is of comparatively recent origin, but seems to have a bright future. The wool averages 56's.

Straight Furrow photo

FIG. 71. *Dorset Horn ewe and lambs*

Straight Furrow photo

FIG. 72. *South Suffolk ram*

Straight Furrow photo

FIG. 73. *South Suffolk ewe and lamb*

SHEARING QUALITIES

The South Suffolk is a good average-shearing sheep, which, although not as good shearing as the straight Suffolk, is a good sheep to shear. Being more dumpy, like the Southdown, he is a bit harder to hold than the straight Suffolk.

Dorset Down

The Dorset Down is one of the larger Down breeds, and is noted for its mutton carcass. As yet it is not widely used in this country, but it has become more popular in recent years. The wool is of quality 50's to 56's.

SHEARING QUALITIES

I have not shorn many sheep of the Dorset Down breed, but those I have encountered have shorn well, and handled well. The wool is good cutting, and fine, and is on a good shapely carcass.

Straight Furrow photo

FIG. 74. *Dorset Down ram*

Straight Furrow photo

FIG. 75. *Hampshire Down ram*

Hampshire Down

The Hampshire Down is another of the Down breeds, noted for carcass quality. The breed is not widely used in New Zealand yet, but several importations have been made in recent years, and several flocks founded. The wool quality averages 50's to 56's.

SHEARING QUALITIES

I have not shorn many of this breed, and those I handled were just average shearing. The Hampshire Down has fine wool on a well-shaped carcass, but the wool on the points does not comb as well as on most of the mutton breeds, being more like that on a Southdown.

Ryeland

The Ryeland is another of the English mutton breeds, in carcass type being a larger edition of the Southdown. Although carrying a slightly heavier fleece than the Southdown, it is mainly a meat-producing sheep, and is very well covered in this respect. Quite a number of farmers, particularly in Canterbury, use Ryeland rams for fat-lamb production, but the breed has not made real progress in New Zealand. The wool is mainly of the Down type of 50's to 56's in quality.

SHEARING QUALITIES

The Ryeland shears quite nicely, and has a fine wool. It is quite a big sheep, with a very thick-set, strong body. Ryelands are not a docile breed, and will often resist the shearer. They are covered with wool well down on the legs and on the head, and the wool on the points does not lift off as it does with some breeds. With their strength and size, and coverage of wool, they are just average shearing, seeming very even, seldom being first-class shearing, and seldom tough.

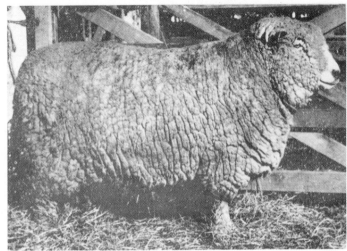

Straight Furrow photo

FIG. 76. *Ryeland ram*

Straight Furrow photo

FIG. 77. *Ryeland ewe*

Scottish Blackface

Although a mountain sheep, the Blackface is the premier breed
of sheep in Great Britain and the most hardy and active of all.
Improvement has not spoiled it; the points that make for
hardiness have been preserved, and artificial food is hostile
to the natural habits of this monarch of the hills and glens of
Scotland. Its constitution enables it to endure the severest of
winter storms, and spring barrenness on mountain land of very
high altitudes. It will travel and climb long distances in search
of food, and when the ground is covered with snow this hardy
sheep will scrape through it to reach the heather and herbs,
while other more delicate breeds would perish.

The breed's economic value is most clearly indicated by its
power to convert the lowest grade of pasture into fine quality
mutton and a type of wool of acknowledged high value in the
making of tweed, carpets and mattresses. Nothing can excel the
texture and flavour of the mutton. The fat differs from that of
most other sheep in the smallness of the globules, which makes
its digestibility very much easier. The early-maturing pro-
pensities of the Blackface have been greatly developed within
the last half century.

The Blackface ewe is a first-class mother and fine milker.
Her lamb from birth will gain $\frac{1}{2}$ lb live weight per day until four
months old, when it leaves her to be slaughtered, or if put on
arable land will thereafter gain $2\frac{1}{2}$ lb per week. After the ewe's
hard spell of five to six years on the hills, she is in great demand
by the lowland farmers for crossing with a Border Leicester
ram to produce the famous 'Greyface' lambs. With her great
milking qualities and hardiness, and the fact that upkeep is
light, she gives an excellent return for the energy and money
invested. In fact, there are vast stretches in the Western world
where the introduction of a Blackface flock would convert
rough and wasting acres into profitable pasturage.

The wool of the Blackface is close-set, of medium length and
texture, free from blue and kemp, and grades at 28's to 40's.

SHEARING QUALITIES

The Blackface is a type of shearing all on its own. It is the
strongest sheep I have shorn, and with its sharp horns on an

active head a shearer can come in for a bruising time. When one gets used to them they are fast shearing as the wool is coarse and open, there is very little wool underneath, and no wool on the points.

One of the big problems in shearing is the length of the wool (sometimes up to 12 inches), which tends to pull the skin out thereby making it easy to cut the sheep. The wool also wraps around the body of the sheep when cut off, and if a shearer does not use the right style he can find the long blow and last hind leg difficult.

I have shorn better breeds, and I have shorn many worse breeds, but now that I have had the opportunity to work out some points of style especially for this breed, I find the Scottish Blackface good shearing. They would, however, be more pleasant without horns.

FIG. 78. *Scottish Blackface ram*

Welsh Mountain

The climate of Wales, particularly of the mountains, is characterised by heavy rainfall—often exceeding 100 inches per annum—accompanied by cold winds, and during the winter months precipitation either in the form of rain, snow or mountain mists may be almost continuous for weeks together. Under such conditions the hardy Welsh Mountain sheep thrive, and the majority never taste any food other than grass, usually of poor quality.

The wool is of moderate length, fine and close, grading at 54's to 56's, and affords the maximum protection to the animal without hampering its movements on rocks or in scrub. The average weight of fleece in a ewe flock kept on a high mountain is from 3 to $3\frac{1}{2}$ lb, but on lower ground it may be as much as 4 lb. The fleeces of rams are considerably heavier. A certain amount of kemp in the wool is favoured by some breeders, who associate it with hardiness, though with careful selection it can undoubtedly be eradicated or reduced to a minimum.

The carcass is free from wasteful fat, and the lean mutton is close in texture, rich in colour, and of unsurpassed flavour. Moreover, a relatively large proportion of the weight of the carcass is found in the more valuable joints. The peculiar conformation, in particular the light shoulders and neck, the powerful well-developed hindquarters, and the strong muscular back, is closely associated with the activity necessary for its existence on its steep mountain grazings.

The Welsh Mountain is the smallest sheep kept in Britain on a commercial scale, and the average carcass of a fat lamb reared on typical mountain land and fattened during autumn on lowland grass does not much exceed 30 lb at the age of nine or ten months. This is due to the poor quality of the mountain grazings. When kept on good low ground the sheep naturally attain a greater size and yield a heavier fleece.

SHEARING QUALITIES

These sheep are fast shearing, but although small they are very active, and it takes plenty of practice to get into top speed on them. The wool shows a lot of 'kemp' with practically no grease. (It is common to wash these sheep before shearing as the washed wool brings a good premium and very little weight is lost by working.) They are dry cutting and the long tails are a handicap. However, being very bare in the points and light in the body, I consider this breed one of the fastest shearing sheep in the world.

(Sheep similar to those shown in the illustration (Fig. 79), i.e. adult Welsh Mountain ewes, unwashed and untouched by blades or machines prior to shearing, were used for my world record of 559 sheep shorn in nine hours at Llangurig, Wales, on 15th July 1960.)

FIG. 79. *Welsh Mountain ewes and lambs*

Clun Forest

This is the ideal general-purpose sheep. Originating on the hilly ranges on the borders of Shropshire, flocks of Clun Forest are now distributed widely throughout England and Wales. Although it is at home on the hills and moorlands, and will fend and forage for itself like a mountain breed, at the same time it is equally suited to more intensive management, and probably no sheep is better fitted to convert good pastures into lean meat. For crop feeding, the wether lambs are much sought after, and feeders have found Clun sheep equal to the Down cross on the Clun, particularly for quick maturity.

The Clun has a clean, open, dark brown face, and the top of the head is nicely covered with white wool. The ears are not too long and are carried high, giving the sheep an alert appearance. Great importance is attached to a strong shoulder, without coarseness, followed by a good spring of rib. This conformation gives rise to a full-bodied sheep with a capacity for foraging, for which the breed is noted. The hind legs are strong and set well apart, with good, clean hocks, giving the sheep an excellent carriage. These hocks are set low down to allow room for a good second thigh, and a full leg of mutton. The fleece is fine and of first quality, grading at 56's to 58's, set close to the skin, and of medium length. Clips are of good weight, ewes averaging 6 lb.

In size the Clun is relatively large. Twin lambs have no difficulty in attaining 45 lb. dressed carcass weight at four months. Older lambs easily reach 55 lb., and wethers carried on to the end of the year reach 80 lb. The meat is of excellent quality and never too fat; in fact, lean fleshing is a definite characteristic of the breed.

FIG. 80. *Clun Forest ram*

SHEARING QUALITIES

The Clun Forest is an average shearing sheep. It is good natured and handles well. Points are reasonably bare and there is not much trimming for the shearer on this sheep. It is inclined to be dry-cutting, and does not show a lot of grease in the wool. For this reason Cluns are never as good shearing as they look.

I was pleased to be able to use them for demonstrations in England, as they always showed a good well-shaped carcass when shorn.

APPENDIX

Shearing Terms

'ALL ON THE BOARD'. The call given when the last sheep of the mob are all in the catching pens.

BARE-BELLY. A sheep with all the wool scraped or dropped off its belly.

BASH. A wild uncontrolled blow by a shearer.

BASHER. A shearer who uses the above type of blows.

BOGGI (pronounced bog-eye). A shearer's handpiece. This term originated in Australia, the early models being shaped like the Boggi lizard. To me it also suggests another meaning, i.e. to keep the handpiece bogged up cutting wool.

BOOTLACE. A long thin strip of skin cut off a sheep, seen mostly on the last side. It usually comes from wrinkled skin.

BOSS'S BOOTS. This means what it says: it is the footwear a shearer watches out for.

BOW-YANGS. Straps or string under the shearer's knees around the trousers.

BROOMIE. The broom hand on the board.

CATCH. The last sheep of a mob. Very often this sheep produces one extra for the tally of the shearer who shears well enough to catch it first. In a tough mob, shearers will acclaim the 'catch' as 'the one we've been looking for all day'.

CHINAMAN. A lock of wool missed by the shearer and left unshorn on the sheep's rump. The shearer does not see it until the sheep has stood up and got out of the porthole.

COBBLER. A hard, difficult-shearing sheep.

COT. A sheep with matted wool.

CRANKY. The man who cranks the engine of an engine-driven plant.

CUT OUT. Finished; the end of the mob. This call is made to the tally clerk before the new mob is started.

DEUCER. A shearer who regularly shears or tops the tally of two hundred sheep a day.

DRUMMER. The slowest shearer on the board. He shears at the bottom of the board, the furthest away from the wool table.

EXPERT. The man who grinds the gear, and keeps the handpieces and plant in order. This man is often a retired shearer.

FLEECE-O. The man who picks up fleeces and who is sometimes abused by shearers.

FLYER. A very fast-shearing sheep; or the best in the mob.

GASPER. A sheep gasping for breath while being shorn. The trouble is usually caused by grass coming up into the sheep's throat.

GONG. Usually a piece of metal, hanging up by the wool table, that is struck to signify the close and start of a run.

GRIND. To sharpen combs and cutters.

GUN. A shearer who has reached the very top, and who has outstanding ability.

KICKER. A sheep that keeps on struggling and kicking while being shorn.

MOCCASIN. A shearer's correct footwear.

PINK 'EM. To make a very good or better than average job of a sheep. Shearers sometimes call this a 'special cut'.

REP. Shearers' voted representative who speaks for all the shearers in the gang in regard to complaints, decisions, etc.

RINGER. The fastest shearer on the board. He usually shears on the first stand next to the wool table. Any rival shearer must shear a bigger tally for three consecutive days before he can take the ringer's stand.

ROUGH 'EM. The opposite to 'pink 'em', and meaning rough shearing and a bad job of the sheep.

ROUSIE. This is the poor old rouseabout who has a great variety of jobs. Australians sometimes call this hand a 'blue tongue'.

RUN. The shearing time worked between official stops, smokos, or meals.

SANDY BACK. A sheep with sand, grit or dirt in the back wool.

SHEEPO. The man who fills the catching pens and works sheep in the shed. Shearers give the call of 'sheepo' when they have caught the last sheep in the catching pen, and to signify that the pen is empty.

SHI-ACK. The usual shed banter, from which none are excused. All in the shed from the boss to the 'rousie' can be the target of good-natured teasing. This does much to make a hard job a lot easier, and all should enter into this woolshed spirit.

'SIXTY-NINE'. The call made to let the shearers and hands know that ladies and visitors are entering the shed, and to give them a chance to put on their best job and be on their best behaviour. Visitors, of course, take this call of 'sixty-nine' to be just a sheep number.

SMOKO. The much looked-for morning and afternoon tea.

SNOB. The last sheep in the pen.

SNOWED IN. Describing the state of affairs when the shearers are ahead of the wool table, so that wool is lying around the floor waiting to go on the wool table. With a good run shearers can sometimes snow the pressers in. Shearers will be seen at their best when they get ahead of the rest of the shed. They enjoy this snowing in, and usually they make the wool flow faster than ever.

SWEAT RAG. A towel used to wipe sweat off in the runs.

TALLY. The number of sheep a shearer does in a day.

TAR-BOY. The hand who walks the board where sheep are subject to the fly and who puts a smear of tar on the cuts made on sheep.

TASSEL. Greasy locks of wool left under the legs and brisket.

TWO-CUT. This happens where the staple of wool is cut above its base on the sheep, making it necessary for the handpiece to come back and trim off a short piece of wool that has been cut before. Such short pieces are of little value.

UP IN THE AIR. Said when a shearer is flashing his handpiece around in the air while shearing, cutting more air than wool.

WHITE-WASHING. Shearing of young lambs, from whom little wool is taken.

'WOOL AWAY'. When the fleece has not been cleared off the board by the 'fleece-o', and is in the shearer's way, a shearer will make this call.

So may I also, having come to the end of this textbook on shearing, say to all readers

WOOL AWAY!

Index

Recommended Reading and Other Sources

Gilfillan, Archer B. *Sheep*. Boston: Little, Brown, 1936, pages 130-135.

Morrell, L. A. *American Shepherd*. New York: Harper & Bros., 1845, pages 177-182.

Randall, Henry S. *Practical Shepherd*. Rochester, New York: D.D.T. Moore, 1863, pages 170-173.

Towne, Charles W. and E. N. Wentworth. *Shepherd's Empire*. Norman, Oklahoma: University of Oklahoma Press, 1945, pages 298-302.

Wentworth, Edward N. *America's Sheep Trials: History and Personalities*. Ames, Iowa: Iowa State College Press, 1948, pages 76; 416-427.

Von Bergen, Werner. *Wool Handbook,* 3rd ed., Vol. I. New York: John Wiley & Sons, 1963.

Alexander, P. and Hudson, R. F. *Wool: Its Chemistry and Physics*. New York: Van Nostrand Reinhold, 1954.

The Shepherd Magazine, Sheffield Massachusetts. For current information on sheep growing in the United States.

Sheep Shearers' Merchandise and Commission Company, P.O. Box #3005, 14 West Platinum Street, Butte, Montana 59701. Principle supplier of shearing equipment and supplies in the United States.

The reader is encouraged to visit the Merrimak Valley Textile Museum, Massachusetts Avenue, North Andover, Massachusetts, for a very thoroughly documented historical presentation of the American wool industry. The museum contains not only an unusually complete collection of machinery having to do with wool and wool manufacture in the United States, but also a library of several thousand volumes and a highly skilled staff.